AGRICULTURE ISSUES AND POLICIES

GENETICALLY ENGINEERED CROPS AND PROPOSED OVERSIGHT

AGRICULTURE ISSUES AND POLICIES

Additional books in this series can be found on Nova's website under the Series tab.

Additional E-books in this series can be found on Nova's website under the E-book tab.

AGRICULTURE ISSUES AND POLICIES

GENETICALLY ENGINEERED CROPS AND PROPOSED OVERSIGHT

BRANDON C. VAUGHN
EDITOR

Nova
Nova Science Publishers, Inc.
New York

For permission to use material from this book please contact us:
Telephone 631-231-7269; Fax 631-231-8175
Web Site: http://www.novapublishers.com

LIBRARY OF CONGRESS CATALOGING-IN-PUBLICATION DATA

Genetically engineered crops and proposed oversight / editor: Brandon C. Vaughn.
p. cm.
Includes index.
ISBN 978-1-60876-067-1 (hardcover)
1. Crops--Genetic engineering--United States--History. 2. Crops--Genetic engineering--Government policy--United States I. Vaughn, Brandon C. II. Fernandez-Cornejo, Jorge. First decade of genetically engineered crops in the United States. III. United States. General Accounting Office. Genetically engineered crops.
SB123.57.G4784 2010
631.5'2330973--dc22

2010012864

Published by Nova Science Publishers, Inc. † New York

CONTENTS

PREFACE

The genetic engineering of agricultural crops is seen as both promising and controversial, with potentially significant implications for the United States and other countries' food security and economic well-being, the environment, and international relations and trade. Proponents cite the potential for enhanced crop yields; more environmentally friendly food production; more nutritious foods; and the increased use of plants to inexpensively produce pharmaceutical compounds, such as human or veterinary drugs, or industrial compounds, such as substances used in paper production or detergent manufacturing. This book examines genetically engineered crops ten years after the first generation of major crops became commercially available and explores issues relating to their economic and environmental impacts as well as consumer acceptance.

Chapter 1 - Genetically engineered (GE) crops—including crops engineered to resist pests or tolerate herbicides—are widespread in the United States and around the world. Taking direction from the 1986 *Coordinated Framework for Regulation of Biotechnology*, the U.S. Department of Agriculture (USDA), Environmental Protection Agency (EPA), and Food and Drug Administration (FDA) regulate GE crops to ensure that they are safe. The unauthorized mixing of some GE crops with non-GE crops has caused controversy and financial harm. GAO examined (1) unautho- rized releases of GE crops, (2) coordination among the three agencies, and (3) additional actions they have proposed to improve oversight. GAO gathered data from agencies and stakeholders; used criteria from prior GAO work to assess coordination; and reviewed agency proposals.

Chapter 2 - Ten years after the first generation of genetically engineered (GE) varieties became commercially available, adoption of these varieties by

U.S. farmers is widespread for major crops. Driven by farmers' expectations of higher yields, savings in management time, and lower pesticide costs, the adoption of corn, soybean, and cotton GE varieties has increased rapidly. Despite the benefits, however, environmental and consumer concerns may have limited acceptance of GE crops, particularly in Europe. This chapterfocuses on GE crops and their adoption in the United States over the past 10 years. It examines the three major stakeholders of agricultural biotechnology and finds that (1) the pace of R&D activity by producers of GE seed (the seed firms and technology providers) has been rapid, (2) farmers have adopted some GE varieties widely and at a rapid rate and benefited from such adoption, and (3) the level of consumer concerns about foods that contain GE ingredients varies by country, with European consumers being most concerned.

In: Genetically Engineered Crops and Proposed... ISBN: 978-1-60876-067-1
Editors: Brandon C. Vaughn

Chapter 1

GENETICALLY ENGINEERED CROPS: AGENCIES ARE PROPOSING CHANGES TO IMPROVE OVERSIGHT, BUT COULD TAKE ADDITIONAL STEPS TO ENHANCE COORDINATION AND MONITORING

United States Government Accountability Office

WHY GAO DID THIS STUDY

Genetically engineered (GE) crops—including crops engineered to resist pests or tolerate herbicides—are widespread in the United States and around the world. Taking direction from the 1986 *Coordinated Framework for Regulation of Biotechnology*, the U.S. Department of Agriculture (USDA), Environmental Protection Agency (EPA), and Food and Drug Administration (FDA) regulate GE crops to ensure that they are safe. The unauthorized mixing of some GE crops with non-GE crops has caused controversy and financial harm. GAO examined (1) unautho- rized releases of GE crops, (2) coordination among the three agencies, and (3) additional actions they have proposed to improve oversight. GAO gathered data from agencies and stakeholders; used criteria from prior GAO work to assess coordination; and reviewed agency proposals.

WHAT GAO RECOMMENDS

GAO recommends that (1) FDA make public the results of its early food safety assessments of GE crops; (2) USDA and FDA develop an agreement to share information on GE crops with traits that, if released into the food or feed supply, could cause health concerns; and (3) USDA, EPA, and FDA develop a risk-based strategy for monitoring the widespread use of marketed GE crops. FDA agreed with the first recommendation, and, with USDA, agreed in part with the second. The agencies agreed in part with the third recommendation. We stand by the recommendations.

WHAT GAO FOUND

Unauthorized releases of GE crops into food, animal feed, or the environment beyond farm fields have occurred, and it is likely that such incidents will occur again. While there is no evidence that the six known releases into the food or feed supply or into crops meant for the food or feed supply affected human or animal health, some resulted in lost trade opportunities. Moreover, the total number of unauthorized releases into the environment is unknown. USDA and EPA have the authority to inspect fields in which GE crops are tested, but crop developers have detected most violations. USDA and EPA have taken enforcement actions in response to violations, ranging from warning letters to significant penalties. The agencies have used lessons learned from unauthorized releases to make regulatory and policy changes. For example, USDA increased inspections of field trial sites for GE crops producing pharmaceutical compounds; EPA discontinued a policy under which a GE crop containing a pesticidal agent could be approved for animal feed, but not for food; and FDA established a voluntary early food safety evaluation program for certain GE crops intended for food use to help mitigate the impact should unauthorized releases occur during field trials, although it has not made these evaluations available to the public.

USDA, EPA, and FDA routinely coordinate their oversight and regulation of GE crops in many respects, but could improve their efforts. Specifically, USDA and FDA do not have a formal method for sharing information that could enhance FDA's voluntary early food safety review for certain GE crops in the field trial stage and support USDA's oversight. Also, the three agencies do not have a coordinated program for monitoring the use of marketed GE crops to determine whether the spread of genetic traits is causing undesirable effects on the environment, non-GE segments of agriculture, or food safety, as recommended by the National Research Council and others.

USDA, EPA, and FDA have proposed regulatory changes intended to improve their oversight of GE crops. In 2007, USDA assessed a wide array of regulatory alternatives that could redefine, on the basis of risk, which GE crops it regulates and how it will respond to unauthorized releases. USDA's fiscal year 2009 budget request also seeks funding for a voluntary system to help GE crop developers employ best management practices to reduce the risk of unauthorized releases. Furthermore, the 2008 Farm Bill required USDA to take actions on lessons learned from its investigation of an unauthorized release of GE rice. EPA has proposed several changes to its regulations for GE crops that produce pesticides, including one change that would distinguish between pesticidal agents produced in GE crops and those applied topically to crops. In 2001, FDA proposed to require that GE food developers notify the agency before marketing their products. However, as of July 2008, FDA had not taken action to finalize the proposed rule, believing its current approach calling for voluntary notice is sufficient.

ABBREVIATIONS

APHIS	Animal and Plant Health Inspection Service
BQMS	Biotechnology Quality Management System
Bt	*Bacillus thuringiensis*
DEIS	draft programmatic environmental impact statement
EPA	Environmental Protection Agency
FDA	Food and Drug Administration
FFDCA	Federal Food, Drug, and Cosmetic Act
FIFRA	Federal Insecticide, Fungicide, and Rodenticide Act
GE	genetic engineering
IES	Investigative and Enforcement Services
OECD	Organisation for Economic Cooperation and Development
OSTP	Office of Science and Technology Policy
PPA	Plant Protection Act
PVCP	plant virus coat protein
USDA	U.S. Department of Agriculture

November 5, 2008

The Honorable Tom Harkin
Chairman
The Honorable Saxby Chambliss
Ranking Member
Committee on Agriculture, Nutrition, and Forestry
United States Senate

The genetic engineering of agricultural crops is seen as both promising and controversial, with potentially significant implications for the United States' and other countries' food security and economic well-being, the environment, and international relations and trade. Proponents cite the potential for enhanced crop yields; more environmentally friendly food production; more nutritious foods; and the increased use of plants to inexpensively produce pharmaceutical compounds, such as human or veterinary drugs, or industrial compounds, such as substances used in paper production or detergent manufacturing. Opponents argue that not enough is known about the safety of genetically engineered (GE) crops and food, and that they should be more rigorously controlled than conventional alternatives. This debate has been exacerbated by several well-publicized cases of unauthorized release of GE crops into the food supply. For example, in August 2006, the U.S. Department of Agriculture (USDA) announced that trace amounts of a regulated variety of GE rice had been commingled with supplies of conventional rice. This announcement led several U.S. trading partners to refuse U.S. rice exports, potentially disrupting the $1.3 billion U.S. rice export market and leading to financial losses for U.S. farmers and exporters. Furthermore, there also is concern that genetic traits could spread from crops into the environment with unintended consequences for plants and animals. This debate may intensify in the future as genetic modifications to crops become more complex, and as pressures build to increase agricultural yields to meet the growing demand for food and biofuel.

Currently, the United States accounts for about 50 percent of the GE crops planted globally. In 2008, GE varieties accounted for about 80 percent of the corn, 92 percent of the soybeans, and 86 percent of the cotton planted in the United States. In 2005, GE varieties accounted for about 93 percent of the canola. To date, the most common characteristics, or traits, engineered into these crops have been resistance to insect pests and the ability to tolerate specific herbicides. The global value of GE seeds sold in 2007 was estimated at $6.9 billion. Food industry sources indicate that over 70 percent of processed foods sold in the United States contain ingredients and oils from GE crops. Increasingly, some countries— including Argentina, Brazil, Canada, and India—have embraced GE crops and food to, among other things, increase yields. Other countries— including many in the European Union and some in Africa—have resisted GE crops and food, citing safety and economic concerns.

Three federal agencies have primary responsibility for regulating GE crops and food in the United States: USDA, the Environmental Protection Agency (EPA), and the Food and Drug Administration (FDA). USDA is responsible for assessing whether GE crops pose a risk as "plant pests" that could directly or indirectly harm plants. To accomplish this, USDA regulates the importation, interstate movement, and release of GE crops into the environment, the latter of which could occur when a developer tests the crop in a field trial. USDA may, upon finding a GE crop does not pose a potential plant pest risk, grant a petition to extend "nonregulated" status to the crop, meaning that it can be moved or released without agency oversight. USDA also has the authority to regulate GE plants as noxious weeds; a noxious weed is any plant or plant product that can injure or cause damage to, among other things, crops, livestock, interests of agriculture, public health, or the environment. EPA is responsible for regulating all pesticides, including those produced by plants that have been genetically modified to protect themselves from insects, bacteria, and viruses. USDA and, to a lesser extent, EPA exercise oversight of the thousands of field trials in which developers have tested new varieties of GE plants since 1987. FDA has primary responsibility for ensuring the safety of most of the nation's food supply and encourages companies to voluntarily submit safety data on a new food or feed derived from GE crops before it is marketed. The President's Office of Science and Technology Policy (OSTP) published the final version of the *Coordinated Framework for Regulation of Biotechnology (Coordinated Framework)* in 1986. This document outlines the federal government's policy for ensuring the safety of GE organisms, including relevant laws and definitions. It was developed in response to concerns that products resulting from genetic engineering might pose greater risks than those resulting from traditional breeding techniques.

In this context, you asked us to examine (1) unauthorized releases of GE crops into the food or feed supply, or the environment; (2) the degree of coordination among the three key agencies that regulate GE crops under the 1986 *Coordinated Framework*—USDA, EPA, and FDA; and (3) additional actions these agencies have proposed to improve the oversight of GE crops and reduce the potential for unauthorized releases.

In conducting this work, we spoke with and reviewed documents provided by officials at USDA, EPA, and FDA as well as OSTP, which is charged with coordinating federal government policy on biotechnology. We also reviewed scientific and technical studies and other literature and spoke with officials in academia, private industry, and consumer groups. We reviewed applicable laws and regulations as well as available public comments on several agency-proposed GE regulations or initiatives as of October 2008. In addition, we reviewed information on all known unauthorized releases of GE crops into the food or feed supply as of September 2008, and on potentially unauthorized releases of GE crops into the environment for the period of January 2003 through August 2007. We assessed the agencies' coordination efforts, using criteria that we have developed in prior work on agency collaboration and coordination.[1] We did not assess the federal regulation of GE animals. Furthermore, we did not assess U.S. efforts to reduce barriers to international trade in GE agricultural commodities. A more detailed description of our objectives, scope and methodology is presented in appendix I. We conducted this performance audit from July 2007 to November 2008 in accordance with generally accepted government auditing standards. These standards require that we plan and perform our audit to obtain sufficient, appropriate evidence to provide a reasonable basis for our findings and conclusions based on our audit objectives. We believe that the evidence obtained provides this reasonable basis.

RESULTS IN BRIEF

Federal agencies have documented six unauthorized releases of GE crops into the food and feed supply or into crops meant for the food or feed supply and additional releases into the environment, as of September 2008, and the ease with which genetic material from crops can be spread makes future releases likely. While the agencies maintain that there is no evidence that any of the known releases have adversely affected human or animal health or the environment, several releases resulted in food recalls or lost trade opportunities that caused financial losses. Moreover, the actual number of unauthorized releases is unknown. Specifically, while USDA and EPA regulations subject crop developers to periodic inspections by federal or state personnel to ensure that developers have taken adequate measures to isolate regulated GE crops from other crops, USDA does not have the resources to inspect all sites, and EPA and the states have not made inspections a priority. In most cases, crop developers have self-reported known unauthorized releases and other violations of regulations. USDA and EPA have taken enforcement actions—ranging from issuing warning letters to assessing significant financial penalties— against GE crop developers who violated regulations. USDA, EPA, and FDA have also taken steps in response to these incidents to reduce the potential for future unauthorized releases and to mitigate the impact of any releases. For example, USDA has increased the frequency of inspections of field trial sites for GE crops producing pharmaceutical and industrial compounds; EPA has discontinued a policy under which a GE crop containing a pesticidal agent could be approved for animal feed, but not for food; and FDA has established a voluntary early food safety evaluation of GE crops that might pose a new risk to help mitigate the impact of unauthorized releases, although FDA has not yet fulfilled a commitment to publish the results of those evaluations.

As called for by the *Coordinated Framework* and measured against other established criteria, the three federal agencies routinely work together to regulate GE crops. For example, the agencies have agreed on their respective roles and responsibilities and developed mechanisms for making policy decisions, sharing information, and responding to incidents. However, the agencies could enhance their coordination by leveraging resources and developing mechanisms to monitor and evaluate results. For example, USDA and FDA do not have a formal method for sharing information that could enhance FDA's voluntary early food safety evaluation of certain GE crops in the field trial stage and USDA's oversight of those field trials. Sharing such information could better leverage resources to address food safety issues for GE crops at the field trial stage. In addition, USDA, EPA, and FDA do not have a coordinated program for monitoring and evaluating the use of marketed GE crops to determine whether they are causing (1) undesirable effects to the environment or economic harm to non-GE segments of agriculture through the unintentional spread of GE traits or (2) food safety concerns, such as the unintentional introduction of pharmaceutical or industrial compounds into the food supply. Several organizations, such as the National Research Council, have made such recommendations regarding the monitoring of GE crops.

USDA, EPA, and FDA have proposed several regulatory changes intended to improve the oversight of GE crops and reduce the potential for unauthorized release. For example, in July 2007, USDA released a draft programmatic environmental impact statement (DEIS) that assessed proposals to modify many aspects of how the agency regulates GE crops, such as how it will respond to the unauthorized release of low levels of GE crops and how it will address the food safety risks posed by GE crops that produce pharmaceutical or industrial compounds when setting requirements for field trials. In October 2008, USDA released for public comment its proposed amendments to those regulations. In addition, USDA's fiscal year 2009 budget request seeks funding to establish a voluntary system to encourage GE crop developers to employ best management practices for field trials and the handling of regulated materials, including third-party audits of their field trial plans and records. The Food, Conservation, and Energy Act of 2008 (2008 Farm Bill) directs USDA to consider regulatory and procedural changes based on the agency's *Lessons Learned and Revisions Under Consideration for APHIS' Biotechnology Framework*, a document resulting from lessons learned from its investigation of the unauthorized release of GE rice into the food supply in 2006, as well as from its years of regulatory experience, and to take action to, among other things, enhance the availability of genetic samples from developers and the quality and completeness of records by developers. For its part, EPA is working on three proposed changes to regulations, including one that would make a distinction between pesticidal agents produced in GE crops and pesticides made from chemicals that are applied topically to crops, noting that currently approved GE-based pesticides are less toxic and, therefore, generally present less risk. FDA proposed in 2001 to require—rather than to encourage, as it does now—developers of GE food products to consult with the agency about the safety of the food before it is marketed. However, as of July 2008, FDA had not taken action to finalize the proposed rule. FDA officials told us that such a rule may no longer be needed because the voluntary consultation process is working well and fully protects the public health.

To ensure that the federal government addresses emerging risks associated with new developments in GE crops, we are recommending that FDA post on its Web site the results of its early food safety evaluations, and that USDA and FDA develop a formal agreement to share information concerning GE crops with novel genetic traits that could cause, or are likely to cause, health concerns if unintentionally released into the food or feed supply. We are also recommending that USDA, EPA, and FDA develop a coordinated strategy for monitoring the marketed use of GE crops for unintended consequences to the environment, non-GE segments of agriculture, or food safety.

In commenting on a draft of this report, USDA, EPA, and FDA generally agreed with the report's findings. On the first recommendation, FDA said it intends to make every effort to fulfill its commitment to post to its Web site the results of completed and future early food safety evaluations. However, FDA also said that activities of greater public health priority have been the focus of its limited resources. Nevertheless, we believe that posting the results of these evaluations would be a low-cost way to increase public transparency and mitigate the impact of unintended releases of GE crops. Regarding the second recommendation, USDA and FDA agreed, in part, saying that they would explore the development of a formal agreement for sharing information on GE crops with novel genetic traits. However, they also said that they should focus their resources on issues that present or are likely to present public health concerns, rather than perceived concerns. We modified this recommendation to remove the reference to "perceived health concerns" and instead emphasize that the agreement would cover GE crops that present or are likely to present public health concerns. Concerning the third recommendation, USDA, EPA, and FDA agreed, in part, to the development of a coordinated strategy to do risk-based monitoring of marketed GE crops for unintended consequences. However, USDA emphasized that its current regulations limit it to monitoring only regulated crops that pose a potential plant pest risk; EPA stated that GE crops that produce pesticides do not require any further post-market monitoring; and FDA said post-market monitoring of food and feed derived from GE crops is not necessary and random sampling to detect GE crops producing pharmaceutical or industrial substances in food and feed would present significant technical challenges and greatly affect resources. Nevertheless, the agencies agreed to enter into discussions to develop a coordinated strategy should such monitoring be necessary in the future. Given that in the United States (1) GE crop varieties are grown extensively, (2) most processed foods contain ingredients from GE crops, (3) it is inherently difficult to prevent the spread of plant genetic material in the environment, (4) there may be an increasing use of GE crops to produce an even wider array of pharmaceutical and industrial compounds in the future, and (5) genetic modifications are becoming increasingly complex in response to pressures to increase yields for food and biofuel, we continue to believe the agencies should develop a coordinated strategy for risk-based monitoring of marketed GE crops.

USDA's and FDA's comments are presented in appendixes II and III, respectively. EPA provided its comments orally. EPA and FDA also provided technical comments that we have incorporated as appropriate.

BACKGROUND

Genetic engineering refers to the technology of modifying the genetic makeup of crops, animals, or microorganisms by introducing genes for specific traits. For centuries, people have crossbred related plants or animal species to develop useful new varieties or hybrids with desirable traits, such as better taste or increased productivity. Traditional crossbreeding, however, can be very time-consuming because it may require breeding several generations to obtain a desired trait and breed out numerous unwanted characteristics. Genetic engineering techniques allow for faster development of new crop or livestock varieties, since the genes for a given trait can be readily incorporated into a plant or animal species to produce a new variety incorporating that specific trait. In addition, genetic engineering increases the range of traits available for developing new varieties by allowing genes from totally unrelated species to be incorporated into a particular crop or animal variety.

Seed developers have experimented with engineering a wide variety of traits into plants, including insect resistance; herbicide tolerance; resistance to viruses, bacteria, and fungi; enhanced product quality, such as increased oil content, delayed ripening, and altered color; and other properties, such as increased tolerance to drought or cold. For example, as shown in figure 1, scientists produced insect-resistant plants by identifying a gene responsible for insect resistance in an organism, isolating and copying the gene, and then inserting the gene into the target plant's DNA.

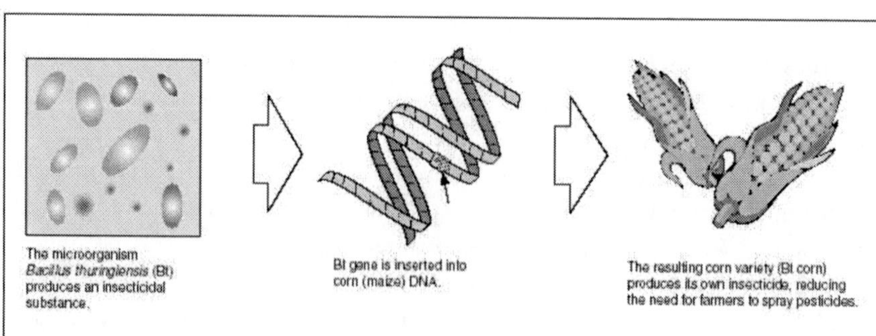

Sources: GAO and Art Explosion (clip art).

Figure 1. Use of Biotechnology to Create a Pest-Resistant Plant

Table 1. Key Statutes Relevant to the Regulation of GE Plants

Statute	Relevance to the regulation of GE crops
Plant Protection Act[a] (PPA)	Authorizes the Secretary of Agriculture to regulate the importation or movement in interstate commerce of plants and articles, including GE crops, that might introduce or disseminate a plant pest or noxious weed.
Federal Insecticide, Fungicide, and Rodenticide Act (FIFRA)	Authorizes the EPA Administrator to register pesticides and regulate the distribution and use of nonregistered pesticides, which would include those genetically engineered into plants.
Federal Food, Drug, and Cosmetic Act (FFDCA)	Authorizes the Secretary of Health and Human Services (delegated to FDA) to regulate food, animal feed, additives, and human and animal drugs, which would include those derived from biotechnology such as GE crops. In addition, authorizes the Administrator of EPA to establish tolerances or tolerance exemptions for pesticidal chemical residues.
National Environmental Policy Act of 1969	Requires all federal agencies to consider the likely environmental effects of actions they are proposing, and if those actions would significantly affect the environment, provide an environmental impact statement. Such statements could be required for actions related to the regulation of GE crops.

Source: GAO.

[a] In 2000, the Plant Protection Act incorporated many authorities of the Federal Plant Pest Act, the Plant Quarantine Act, and the Federal Noxious Weed Act of 1974 and repealed all but a few provisions of those acts.

In 1986, OSTP published the *Coordinated Framework*, which outlined the regulatory approach; relevant laws; and regulations for, and a definition of, GE organisms. This document states that existing statutes provide a basic network of agency jurisdiction over genetic engineering both for research and products. The statutes most relevant to the regulation of GE crops are shown in table 1, with additional details provided in appendix IV. In 1992, OSTP elaborated on the *Coordinated Framework* with a policy announcement that (1) called for the oversight of GE organisms only when there is evidence of "unreasonable" risk, that is, when the reduction in risk obtained by oversight is greater than the cost of oversight, and (2) expected federal agencies to focus on the characteristics and risks of biotechnology products, not on the process by which these products are created.

Responsibility for implementing the *Coordinated Framework* fell primarily to three agencies—USDA, EPA, and FDA—with USDA designated as the lead agency for plants and animals. Each agency has specific requirements for certain activities with GE crops, and not all three agencies are necessarily involved in overseeing each activity or use of a GE crop. The applicability of these requirements to GE crops depends upon several factors, including the type of trait engineered into the plant and the proposed use of the crop. Specific responsibilities of the agencies are described in the following text.

USDA Oversees the Movement and Environmental Releases of Regulated GE Crops

USDA regulations require persons seeking to import, move interstate, or release into the environment GE crops to first submit a notification to the agency or obtain a permit, depending on the risk that the GE crop poses, with notification being the more administratively streamlined option:

- *Notification*: USDA regulations provide that GE crops may be released into the environment or moved under a notification, rather than with a permit, if they meet the following six criteria.

    1. The GE crop species is not listed in regulation as a noxious weed or considered by the Administrator of USDA's Animal and Plant Health Inspection Service to be a weed for the area of release.
    2. The introduced genetic material is "stably integrated" in the crop's genome.
    3. The function of the introduced genetic material is known and does not result in plant disease.
    4. The introduced genetic material does not cause the production of an infectious entity, produce a substance that is likely to be toxic to nontarget organisms, or produce a product intended for pharmaceutical or industrial use.
    5. The introduced genetic sequences do not pose a significant risk of the creation of a new plant virus.
    6. The crop has not been modified to contain certain genetic material from animal or human pathogens.

    USDA regulations also require that activities conducted under a notification meet certain performance standards. Namely, regulated

GE crops must be handled in such a way that they do not persist in the environment or get mixed with nonregulated plant materials. A general technique for avoiding mixing is to isolate the GE crops from non-GE crops, and USDA has described in guidance documents a number of steps that developers may take, such as bagging or netting the plants to contain the seeds, planting border rows, or using sterile male varieties.

- *Permit*: The USDA permit process is for those GE crops that cannot be introduced under notification, such as plants engineered to produce pharmaceutical or industrial compounds or modified with genetic material that causes the production of an infectious entity or toxic substance. Permits spell out specific requirements for conducting the activity, with the permit conditions for GE crops that produce pharmaceutical or industrial compounds typically being the most restrictive. For example, permit conditions for these types of GE crops require that the fallow zones around field trial sites be larger than for other types of crops, that farmers use dedicated machinery (harvesters or planters) and storage facilities, and that the permit holder implement a training program for its personnel.

Permits or notifications are also required for the interstate movement or importation of regulated GE crops. For example, the requirements relevant to these permits address such matters as the points of origin and destination, packaging, and record keeping.

From fiscal years 1987 through 2007, USDA issued almost 19,000 notifications and almost 4,300 permits for environmental releases, importation, and interstate movement. Over 13,000 of the notifications and permits were for releases into the environment, also known as field trials.[2] A single permit or notification for a field trial may cover more than one location at which a GE crop can be tested. (See app. V for details on the yearly rate at which USDA has issued permits and acknowledged notifications for field trials and on the types of genetic characteristics those trials covered.)

USDA regulations also allow for persons, including GE crop developers, to petition the agency to deregulate a GE crop. If USDA deregulates the crop, it is no longer subject to regulatory control under the Plant Protection Act, unless USDA finds it to be a plant pest or noxious weed on the basis of new data or analysis. Petitioning USDA is the typical route to commercialization, since it allows planting with less restrictive conditions than those imposed by a permit or the notification process. However, according to USDA officials, a GE crop developer could market a product that is still regulated. As of July 2008, USDA had received 113 petitions for deregulation and approved 73. (See app. VI for more details on deregulated and marketed GE crops.)

EPA Regulates Pesticides Produced in GE Crops

EPA is responsible for regulating the genetic materials engineered into a crop to produce pesticides that ward off insects, bacteria, and viruses, as well as the pesticide that the crop ultimately produces (known as a "plant-incorporated protectant," but referred to in this chapteras a "GE pesticide"). As with conventional chemical or biological pesticides, EPA regulates the sale, distribution, and use of GE pesticides, and producers must register them before they are put into commercial use. Since 1995, EPA has registered 29 GE pesticides engineered into 3 crops—corn, cotton, and potatoes—5 of which have since been voluntarily canceled. (See app. VI for more details about EPA's process for registering GE pesticides.)

EPA requires persons seeking to conduct field trials of GE crops containing pesticides on more than 10 cumulative acres to apply for an experimental use permit. These crops generally have shown promise in previous small-scale field trials (less than 10 cumulative acres) regulated by USDA and are potential candidates for future commercialization. To receive a permit, applicants must submit data to EPA on the descriptions and specific results of any appropriate prior testing of the product conducted by the applicants to determine toxicity, effects on the environment, and other matters associated with the GE pesticide. According to EPA, it requires that applicants demonstrate that regulated genetic material will not spread into other plants. In the absence of such a showing, EPA will impose containment measures which may be similar to those that USDA requires to address potential environmental risks. If it can be reasonably expected that the field trial will result in pesticide residues in food or feed, the applicant must submit evidence that a tolerance or tolerance exemption has been established or submit a petition for the establishment of a tolerance or tolerance exemption, or certify that the food or feed is disposed of in a manner that ensures it will not endanger man or the environment.

Although EPA establishes tolerances, FDA, not EPA, is responsible for enforcing tolerances for pesticide residues on foods derived from GE crops. If EPA determines that there is a reasonable certainty that no harm will result from aggregate exposure to the pesticide residue, it may grant an exemption from the requirement for a tolerance. FDA may take enforcement actions if residue of a GE pesticide enters into the food supply without a tolerance or exemption from tolerance. From fiscal years 1997 through 2007, EPA issued 65 experimental use permits for field trials of GE pesticides, or about 6 such permits per year. As of June 1, 2008, there were 8 active permits for GE pesticides, covering about 26,000 acres. According to EPA, it generally considers small-scale field trials to have adequate containment measures if they are conducted under USDA authorization and are in compliance with USDA requirements and meet EPA's requirement that no pesticide residues can be in the food or feed supply unless there is a tolerance or tolerance exemption in place.

FDA Encourages Developers to Consult on Food Safety Issues before Marketing GE Crops, and Regulates GE Pharmaceutical Products

FDA is responsible for ensuring the safety of most of the food supply, with the exception of meat, poultry, and egg products, which are under USDA's authority. FDA established its basic policy regarding the review of GE foods in its 1992 *Statement of Policy: Foods Derived from New Plant Varieties*, which explained that substances introduced into food or feed by way of breeding were potential food additives if they were not generally recognized as safe or if they were pesticides, and described the kinds of assessments FDA expected companies to perform to assure themselves that foods and feeds from new plant varieties were as safe as comparable foods and feeds already on the market, and otherwise did not raise regulatory concerns. In 1995, FDA established its voluntary consultation process, through which companies developing foods and feeds from GE plants voluntarily notify the agency and submit a safety assessment report containing a summary of test data and other information on the foods before they are marketed. The company evaluates, for example, whether the level of allergens, toxins, nutrients, and antinutrients—compounds that inhibit the absorption of nutrients—in the GE food is comparable to the level of these substances in the food's conventional counterpart, and whether the GE food contains any new allergens or toxins. FDA assists the company with questions related to the regulatory status of the food. If FDA has no further questions about the safety of the food or feed, it provides the company with a letter to that effect. Although the consultation process is voluntary, it is FDA's experience that companies do not commercially market their GE crops until they have received this letter. As of July 2008, FDA had completed 72 voluntary consultations on GE crops intended for use in animal feed, human food, or both. FDA does not track which of these GE crops have been marketed; industry data indicate that many have been, but that some are no longer commercially available. (See app. VI for more details about GE crops that developers have marketed.)

For plants engineered for a nonfood use, such as those that produce a pharmaceutical compound, FDA subjects the pharmaceutical product to the drug or biologic review and approval process. In 2002, in collaboration with USDA, FDA published draft guidance to the biotechnology industry that outlined some of the steps industry should take to ensure that regulated products do not become mixed with the food or feed supply and manufacturing information that should be submitted to FDA with applications for marketing approval.

Unauthorized Releases of GE Crops Have Caused Financial Losses, and the Agencies Have Taken Steps to Reduce the Likelihood of Future Releases

Federal agencies have documented the unauthorized releases of regulated GE crops into the food supply and the environment. While none of these releases are known to have affected human or animal health or the environment, some releases into the food supply had substantial financial consequences. Specifically, there have been six known releases of GE crops into the food or feed supply or into crops meant for the food or feed supply; with the first occurring in 2000. While these releases have not harmed human or animal health, several had significant financial consequences, including product recalls and destruction and lost trade opportunities. USDA data indicate that there have been more unauthorized releases of regulated crops into the environment, but the agency says that they have not caused environmental harm. USDA and EPA have taken enforcement actions in response to violations, including several large financial penalties. All three agencies have taken steps in response to known unauthorized releases to reduce the potential for future unauthorized releases or to mitigate their impact.

Known Unauthorized Releases of GE Crops Apparently Have Not Caused Health Effects, but Several Caused Financial Losses

There have been six known incidents of the unauthorized release of regulated GE crops into the food supply or into crops meant for the food supply—four involving GE varieties of corn and the remaining two involving a GE variety of rice. (See table 2.) These incidents apparently have not caused health effects, but several led to financial losses for farmers and exporters. While the specific causes of unauthorized releases vary by incident, from cross-pollination of regulated and conventional crops to the mislabeling of bags of seeds, they highlight the challenges of containing regulated GE crops given the porous nature of biological systems and the potential for human error. (See app. VII for a detailed description of each of these incidents.) According to USDA, large-scale annual field testing of GE crops occasionally results in materials from these trials being detected at low levels in commercial commodities and seeds. Most officials we asked, including representatives from the biotechnology industry, agricultural commodity growers, and consumer advocacy organizations, also told us that future unauthorized releases of low levels of regulated GE material are likely to occur.

Table 2. Summary of the Six Known Unauthorized Releases of Regulated GE Crops into the Food and Feed Supply, 2000-2008

Year	Product	Crop	Trait	Cause	Detection
2000	StarLink	Corn	Insect resistance and herbicide tolerance	Cross-pollination, commingling of corn after harvest	Third-party testing
2002	Prodigene	Corn	Pharmaceutical protein	Cross-pollination and uncontrolled volunteers[a]	USDA insp-ection
2004	Syngenta Bt1 0	Corn	Insect resistance	Misidentified seed	Third-party testing
2006	Liberty Link Rice 601	Rice	Herbicide tolerance	Not determined	Third-party testing
2006	Liberty Link Rice 604	Rice	Herbicide tolerance	Not determined	Third-party testing
2008	Event 32	Corn	Insect resistance	Under investigation	Deve-loper testing

Source: GAO analysis of USDA and EPA data.

[a] "Uncontrolled volunteers" refers to plants from a previous season's field trial that grow on their own without being deliberately planted.

The regulated materials in these six incidents were detected at different points in the food and feed supply. For example, in the StarLink corn incident—a GE corn containing a pesticidal protein that was approved only for animal feed and not for human food—trace amounts of the pesticidal protein were detected in consumer products, such as taco shells and corn bread. The presence of the pesticidal protein in human food rendered it adulterated. Therefore, FDA requested food processors to recall potentially affected food products. In the Prodigene corn incident, USDA discovered that the regulated crop had been mistakenly harvested and commingled with soybeans in a grain silo. USDA ordered the soybeans and GE corn destroyed before they were sold commercially.

Table 3. USDA Data on Incidents from January 2003 through August 2007

Total number of permits and notifications	Number of violations or potential violations	USDA categories of violations that could indicate a release to the environment	Number of violations or potential violations identified by USDA[a]
6,983	712	Persistence in the environment[b]	7
		Production and/or persistence of progeny (offspring)	4
		Animal-related release, incursion, destruction, or consumption	33
		Weather-related release, incursion, or destruction	17
		Movement of propagules into the environment[c]	16
		Isolation distance or other flower control Insufficiency	21

Source: USDA.

[a]An incident may involve more than one violation or potential violation.

[b]A GE crop that is persistent in the environment is one that produces a sustained population in agricultural or nonagricultural habitats without human intervention.

[c]Propagules are any part of a plant that can be detached from the organism and propagated in order for it to grow into a new plant.

With the exception of Prodigene corn, the regulated material in all of the incidents involved traits of herbicide tolerance and insect resistance familiar to federal regulators. In addition, the regulated materials found in Syngenta Bt10, Liberty Link Rice 601 (LLRICE 601), Liberty Link Rice 604 (LLRICE 604), and Event 32, were very similar to GE material that had already been reviewed by EPA, FDA, or both, and deregulated by USDA. Shortly after each of these four incidents, EPA, FDA, or both, issued statements attesting to the safety of the low-level presence of the regulated GE crops in the food and feed supply.

While USDA, EPA, and FDA have determined that none of these six incidents of unauthorized release harmed human or animal health, some cases led to financial losses, particularly from lost sales to countries that would not accept crops containing the regulated GE varieties. For example, in response to the detection of regulated GE rice in commercial rice supplies in the United States in 2006, several of the leading importers of U. S. rice either banned the import of certain varieties of rice imports or imposed new testing requirements on rice traders. However, it is difficult to quantify the financial losses resulting from these unauthorized releases because many factors may determine the final sale price of commodity agriculture. Of the few estimates available, one by a group of economists estimated that the StarLink incident resulted in $26 million to $288 million in lost revenue for producers in market year 2000/2001. (According to USDA, U.S. cash receipts for corn totaled about $15.2 billion in 2000.) Similarly, a separate study by university economists estimated that the presence of StarLink in the food supply in 2000 caused a 6.8 percent drop in the price of corn, lasting for 1 year. More recently, an environmental advocacy group estimated that the worldwide costs resulting from the LLRICE incidents, including the costs associated with the loss of export markets, seed testing, elevator cleaning, and food recalls in countries where the variety of rice had not been approved, ranged from $741 million to $ 1.285 billion.

USDA Says That Unauthorized Releases of GE Crops Have Not Caused Environmental Harm

In addition to known unauthorized releases to the food supply, USDA data indicate there have been other potentially unauthorized releases of GE crops into the environment. However, USDA has concluded that these releases have not caused harm. Most of the reports of such incidents were self-reported by the developers, rather than identified through USDA inspections. In 2007,

USDA analyzed its record of over 700 violations or potential violations that occurred from January 2003 through August 2007 and found 98 that indicated a possible release into the environment, as shown in table 3.

A concern associated with the release of a GE crop into the environment is that its pollen containing its genetic characteristics may spread to wild relatives. This is known as “gene flow.” There is the potential for the traits of insect resistance or herbicide tolerance to transfer to weedy relatives of a crop, which could give the weeds a competitive advantage or require a different herbicide for their control. The turf grass known as “creeping bentgrass” is an example of this concern. The Scotts Company has tested herbicide-resistant creeping bentgrass in the hopes that it can be marketed for use on golf courses and lawns. In 2003, several environmental organizations and individuals filed suit against the Secretary of Agriculture and other officials for, among other things, permitting field tests of GE creeping bentgrass without adequately determining whether the crop was a plant pest that could spread to wild relatives or preparing an environmental impact statement or environmental assessment pursuant to the National Environmental Policy Act. Evidence presented in the case showed that the GE bentgrass at the field test site had pollinated wild relatives. The court found in February 2007 that there was no evidence that USDA considered whether the permitted field tests had the potential to significantly affect the environment when it decided that an environmental impact statement or assessment was not necessary. The court held that USDA could not process future permits without first considering whether the field tests involve either new species or organisms or novel modifications that raise new issues and, if either one exists, whether the field tests likely would significantly affect the quality of the human environment.

Agency Inspections Have Led to Some Enforcement Actions and Penalties

USDA and EPA have the authority to conduct inspections of field trials and other activities, and the agencies do so under their respective regulations to help ensure compliance. USDA does not inspect all field trial sites where GE crops are tested; instead, it uses a risk-based approach to select sites for inspection. In response to violations, USDA has taken enforcement actions, such as issuing enforcement letters and assessing financial penalties. EPA, on the other hand, has delegated primary enforcement authority—including inspection responsibilities—to the states, but, according to EPA, neither the agency nor the states have made these inspections a priority. In response to violations, EPA has assessed several large financial penalties, but otherwise has taken few enforcement actions. However, USDA, EPA, and FDA have taken other actions in response to incidents of the unauthorized release of GE crops to reduce the likelihood of future releases or minimize their impact.

USDA Follows a Risk-Based Approach to Inspect Field Trial Sites

USDA policy is to use a risk-based approach to selecting which field trials covered by permits and notifications it will inspect. The agency's most stringent policy applies to permits for field trials of GE crops engineered to produce pharmaceutical or industrial compounds. For those GE crops, USDA's policy calls for up to 7 inspections of permitted field trials, both during and after the growing season. For permits other than those for pharmaceutical or industrial compounds, USDA's policy is to inspect every permit at least once in each state in which a field trial is done. For example, if a permit allows for 15 field trial sites to be planted in 7 states, at least 1 inspection will be done in each of the 7 states. According to USDA officials in charge of the inspection program, the agency has met the inspection goals for permitted field trials in recent years.

USDA policy does not call for inspecting all field trials done under a less stringent notification. For fiscal years 2005 through 2007, USDA selected for inspection about one-third of the field trials conducted under the notification procedure on the basis of the developer's past compliance record, the size of the field trial, the number of field trial sites covered by the notification, and the type of crop being tested, among other factors. A developer may conduct notification field trials at many sites, but USDA does not necessarily inspect all of those sites.

During inspections, USDA officials check records, make visual or photographic observations, and conduct interviews to determine regulatory compliance, including whether regulated material might have been inadvertently released. However, these officials told us that they do not have the resources to develop methods to conduct genetic testing of the area surrounding a field test site as part of routine inspection to determine with certainty whether regulated genetic material has escaped the control of the biotechnology developer. Instead, USDA relies on biotechnology developers to voluntarily provide them with the genetic testing methodology and representative samples necessary to detect regulated articles when USDA has reason to believe they may have been released from a site. According to USDA officials, to date, developers have been cooperative when asked to provide a testing methodology and representative samples, although doing so is not a requirement of the regulation.

Although USDA's inspection program has detected some violations of regulations, it generally has found a high rate of compliance. Over the 3-year period from fiscal years 2005 through 2007, USDA inspected field trials conducted under 489 permits and found that 18 (about 4 percent) were out of compliance. USDA also found high compliance levels at field trials operated under a notification; it completed 754 inspections over the same period and found 17 instances of noncompliance (about 2 percent). Holders of USDA permits and notifications are required to self-report, and most incidents have been identified by self-report, rather than by inspection.

USDA Has Taken Enforcement Actions in Response to Violations and Has Assessed Financial Penalties

From calendar years 2003 through 2007, USDA's typical enforcement action in response to regulatory violations generally was to issue an order requiring the developer to take corrective action; in a small number of other cases, USDA also obtained a civil monetary penalty from the developers. USDA handled 320 incidents representing violations or potential violations reported during this period. These incidents included those self-reported by the developer and those detected by USDA inspections. USDA resolved more than half of the incidents with an acknowledgment letter or notice indicating that the developer had returned to compliance or that the alleged incident was not, in fact, a violation. The remaining incidents led to guidance letters or notices of noncompliance, warning letters, or referrals to USDA's Investigative and Enforcement Services (IES) (see figure 2).[3]

According to USDA officials, the agency refers serious incidents to IES, and only incidents that have been referred to IES have resulted in fines. We reviewed case files associated with the 17 referrals to IES. Over half were initiated in 2005 to address nationwide noncompliance by the Scotts Company in its development of GE creeping bentgrass. Those violations included allowing GE grasses to form pollen that might have pollinated plants outside of the field trial site, exceeding the allowable acreage in a field trial, missing records for particular field trial sites, allowing unauthorized movement of regulated GE grass to locations outside of the field trial site, and lacking adequate borders around field trial sites. In 2007, the company entered into a consent decision with USDA and agreed to conduct three compliance workshops and pay a $500,000 fine, which is the maximum allowable under the statute.

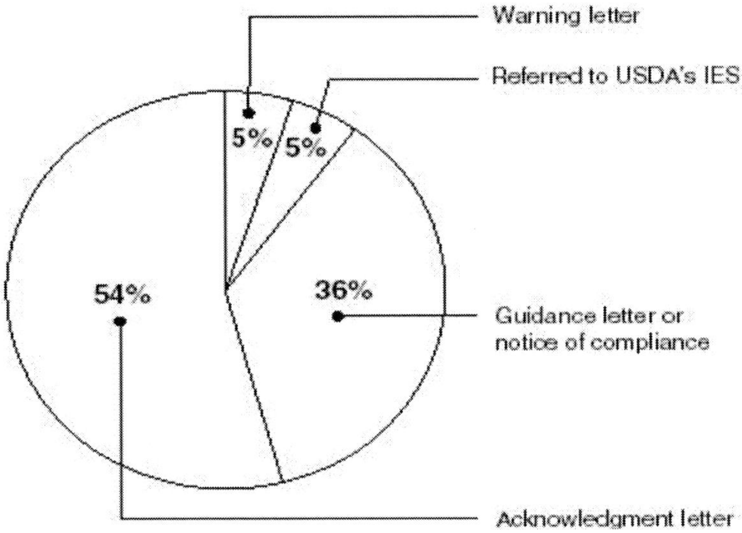

Source: GAO analysis of USDA data.
Note: Starting in 2005, guidance letters were changed to notices of noncompliance.

Figure 2. USDA Enforcement Actions, 2003 through 2007

Also among the referrals to IES was the 2004 Syngenta Bt10 corn incident, for which USDA levied a $375,000 fine. Seven other incidents that IES investigated in 2005 through 2007 resulted in financial penalties ranging from $2,500 to $40,000. The seven violations included failure to list a field trial site for a drought-resistant corn in its permit; report that a storm blew regulated bentgrass outside of the field trial site; and maintain the identity of regulated eucalyptus trees being grown in a field trial. In four of these seven incidents, the violator self-reported the incident.

EPA Has Delegated Its Inspection Responsibilities to the States

EPA has delegated its primary enforcement authority, including responsibility for most inspections of field trials conducted under experimental use permits, to state agencies. However, according to EPA officials, neither EPA nor the states consider monitoring field trial permits for GE pesticides a high priority, partly because all of the GE pesticides currently being grown in field trials have already been evaluated for food, feed, and environmental risks and received a tolerance or a tolerance exemption, indicating they are relatively low risk.[4] EPA does not collect information on how many experimental use permits the states inspected. Furthermore, EPA does not collect detailed information on the results of inspections. EPA can initiate its own investigation when there is reason to believe that an applicant is not meeting permit requirements. EPA officials told us that they exercised this option for two field trials conducted by two companies in Hawaii. In those instances, EPA targeted these field trials for inspection because the permit holders did not have a tolerance for the GE pesticide they were testing. EPA inspectors found permit violations that could have resulted in the unauthorized release of a GE pesticide. EPA officials said that, following these incidents, the agency stopped issuing experimental use permits for field trials of GE pesticides that do not have a tolerance or an exemption from tolerance.

EPA Has Assessed Several Large Financial Penalties, but Has Taken Few Other Enforcement Actions Related to GE Pesticides

EPA has assessed several large financial penalties since it began to regulate GE pesticides in 1986. Otherwise, however, the agency has taken few enforcement actions. As of August 2008, EPA had issued financial penalties on four occasions for violations of pesticide laws and regulations involving GE pesticides, ranging from $8,800 to $1.5 million. Two of these occasions were related to violations of field trial permit conditions in 2002. During the inspections in Hawaii of field trials of GE pesticides being grown without a pesticide tolerance, EPA inspectors found that one permit holder had planted experimental corn in an unapproved location, and that another permit holder did not have an appropriate buffer surrounding the field trial. The permit holders were fined $9,900 and $8,800, respectively. In addition, as part of its settlement, EPA required the first company to perform tests to confirm that the experimental gene grown in the field trial had not been transferred to adjacent fields. In 2003, EPA imposed an additional $72,000 penalty on that company for failing to immediately report to the agency the results of an initial test that suggested that an inadvertent release of an unregistered pesticide had occurred. Subsequent testing conducted by the company revealed that the initial test had been incorrect, but EPA still fined the company for failing to report the initial test results. On the remaining two occasions, EPA issued fines of $165,200 in 1996 and $1.5 million in 2006 in response to separate incidents of the unauthorized sale and distribution of a registered pesticide. The latter fine related to the unauthorized release of Bt10 corn, as we have previously discussed. Other enforcement options available to EPA include calling for the destruction of products, as it did with Bt10 corn, or stopping the sale of a product, as it did in the case of Event 32 corn.

Agencies Have Taken Actions in Response to Incidents of Unauthorized Release to Reduce Their Likelihood or Minimize Their Impact

In response to incidents of unauthorized release, USDA, EPA, and FDA have taken several actions to either reduce the likelihood that regulated crops would be unintentionally released into the food supply or the environment or to minimize the impact of such occurrences. In some cases, these actions were a response to specific incidents. For example, the StarLink corn incident led to two significant policy changes in the way that EPA regulates GE pesticides. First, EPA decided to stop issuing split registrations—in which a product is approved for animal feed but not for human consumption. StarLink had been the first—and the only—GE pesticide to receive a split registration. Second, EPA began requiring developers of GE pesticides receiving a tolerance or an exemption from tolerance to develop a quick-detection method for the modified gene and provide it to EPA as part of the product's registration. In addition, in March 2003, not long after the Prodigene incident, USDA published a request for comments in the *Federal Register* that included a description of more stringent permit conditions for environmental releases of plants that produce pharmaceutical and industrial compounds. USDA also announced that it would increase the number of USDA field trial site inspections, stating that a field test may have five inspections during the growing season and two additional inspections postharvest to look for volunteer plants. In addition, USDA would restrict what can be grown on a test site and fallow zone in the next growing season.

Other actions have been a response to releases in general. In an August 2002 *Federal Register* notice, OSTP articulated three principles regarding field trials of GE crops: (1) the level of confinement under which field tests are conducted should be consistent with the risks posed; (2) if the risk is unacceptable or unknown, field trial confinement requirements should be rigorous to prevent unauthorized releases, and the occurrence of any genes and gene products from those field tests in commercial seed, commodities, and processed food and feed would be prohibited; and (3) in other instances where risks are low, field trial requirements should still minimize unauthorized releases of gene products, but a low level of GE crops in the environment could be found acceptable if available data find that they meet applicable regulatory standards. Following that announcement, USDA, EPA, and FDA published notices concerning their responsibilities regarding field trials and the low-level presence of regulated GE material. Specifically:

- In March 2007, USDA published its current policy for responding to low levels of regulated GE plant materials that may occur in commercial seeds or grain. For example, USDA may determine that remedial action is not necessary when (1) the regulated material is derived from plants that meet all of the criteria to qualify for USDA's notification process and (2) the regulated GE crop is similar to another GE crop that has already been deregulated by USDA. USDA also stated that it could take enforcement action against violators of regulations, even if it decided that no remedial actions were necessary to address the low-level presence of regulated GE material in commerce.
- In May 2007, EPA released guidance for small-scale field testing and the low-level presence of GE pesticides in food. EPA stated if there is any reasonable expectation that residues of the GE pesticide being tested could enter the food supply, even at low levels, all crops affected by such tests must either be destroyed or be kept from the food or feed supply while additional studies using the crop are conducted, or the applicant must obtain a tolerance or tolerance exemption, regardless of the size of the field trial. EPA's policy also noted the FFDCA provision that a food containing pesticide residues may not be moved in interstate commerce without an appropriate tolerance or tolerance exemption. EPA also described methods that developers can use to isolate GE pesticides from the food or feed supply.
- In June 2006, FDA issued guidance recommending that developers of certain GE crops intended for food use, but still in the field trial stage, engage in what the agency called a voluntary early food safety evaluation, whereby developers would consult with FDA about new GE materials produced in these plants before they might inadvertently enter the food supply.[5] If FDA had already reviewed the GE material and had no safety concerns, the agency did not expect developers to participate. FDA has conducted seven such evaluations since 2006. FDA officials said the agency does not use data from USDA's permits database to identify field trials that might be candidates for FDA's early food safety assessments; instead FDA relies on developers for notification. In this guidance, FDA stated that "consistent with confidentiality requirements," it would make the developers' submissions and FDA's responses easily accessible to the public via the Internet. However, FDA has not done so. Agency officials

indicated that they intend to fulfill this commitment to make submissions available online, but FDA has not had the resources to post the submissions.

Routine Interagency Coordination of Programs Occurs, but Opportunities Exist for Further Coordination among the Agencies

USDA, EPA, and FDA have organizational structures and mechanisms in place to coordinate their oversight and regulation of biotechnology, but opportunities exist for further coordination and collaboration among the agencies. Using as criteria practices we have identified in prior work that can enhance and sustain collaboration among federal agencies, we found that agencies have agreed on roles and responsibilities and have established compatible policies, procedures, and other means to operate across agency boundaries. However, the agencies could enhance their coordination by further leveraging resources, developing mechanisms to monitor and evaluate results, and implementing other practices. While we have identified areas for improvement, most of the officials with whom we spoke did not indicate that they had major concerns about the adequacy of interagency coordination, nor did they identify changing the *Coordinated Framework* as a high priority.

Agencies Could Enhance Coordination by Further Leveraging Resources, Developing Mechanisms to Monitor and Evaluate Results, and Implementing Other Practices

GAO has previously identified a set of eight practices that can enhance and sustain collaboration among agencies.[6] Seven of these practices are as follows:

- Defining and articulating a common outcome.
- Agreeing on roles and responsibilities.
- Establishing mutually reinforcing or joint strategies.
- Identifying and addressing needs by leveraging resources.
- Establishing compatible policies, procedures, and other means to operate across agency boundaries.

- Reinforcing agency accountability for collaborative efforts through agency plans and reports.
- Developing mechanisms to monitor, evaluate, and report on the results.

We evaluated the degree of coordination and collaboration among USDA, EPA, and FDA in their oversight of GE crops according to each of these practices.

Defining and Articulating a Common Outcome

The three agencies are working toward the broad common outcome that was originally described in the *Coordinated Framework*. The document sought to achieve a balance between developing regulations adequate to ensure health and environmental safety and maintaining sufficient regulatory flexibility to avoid impeding the growth of the nascent biotechnology industry. To arrive at this outcome, the *Coordinated Framework* attempted to distinguish those organisms that require a certain level of federal review from those that do not. In general, the *Coordinated Framework* and subsequent policy statements from OSTP direct federal agencies to exercise oversight of GE organisms only when there is evidence of unreasonable risk—that is, when the value of the reduction in risk obtained by additional oversight is greater than the additional regulatory costs. Although the types of GE crops that each agency regulates vary, all three agencies have striven to achieve this common outcome through the development of risk-based regulatory systems. For example, USDA's two-tiered permit system, which we previously described, allows for GE crops that present less risk to be eligible for the more streamlined notification procedure, rather than a permit. USDA and EPA have begun other initiatives—which we discuss later in this report—intended to make their oversight of GE crops more risk-based. Similarly, because FDA considers most transferred genetic material to be generally recognized as safe, it does not expect transferred genetic material to be subject to its food additive regulation.[7]

Agreeing on Roles and Responsibilities

The agencies have generally agreed on their roles and responsibilities as they are outlined in the *Coordinated Framework*, which states that existing laws provide the basic network of agency jurisdiction and that jurisdiction over a GE product should be determined by its use. When these responsibilities overlap, the *Coordinated Framework* establishes a lead agency. When incidents of unauthorized release have occurred, the three federal agencies have taken actions related to their roles and responsibilities to protect health and environmental safety. For example, after the most recent unauthorized release, which involved a regulated GE corn known as Event 32, USDA issued emergency action notifications for the unauthorized movement of a regulated article, and EPA issued a "Stop Sale Order" to the developer of the GE corn because it is illegal to distribute any pesticide not registered under FIFRA. The three agencies also issued a joint statement in which USDA concluded that Event 32 poses no plant pest or environmental concerns; EPA determined that the pesticidal material produced by Event 32 is identical to that found in an approved GE pesticide and, therefore, it is covered by an existing tolerance exemption; and FDA concluded there were no food or feed safety concerns.

Establishing Mutually Reinforcing or Joint Strategies

The three agencies have taken steps that establish mutually reinforcing strategies. For example, in 2002, OSTP proposed a mutually reinforcing joint strategy to address how agencies should respond to a low-level presence of regulated GE material in the environment or commercial agriculture. Following OSTP's proposal, USDA issued a notice in the *Federal Register* stating its policy for responding to any occurrences of low-level presence of regulated GE crop materials; EPA released guidance for small-scale field testing and the low-level presence in food of GE pesticides; and FDA issued guidance to recommend that developers of certain GE crops intended for food use but still in the field test stage engage in what it called a voluntary early food safety evaluation, as we have previously described. However, OSTP's proposal was limited in scope to GE crops intended for food or feed use; pharmaceutical and industrial compounds were not a part of the joint strategy. Food plants, such as corn and soybeans, are used to produce these compounds.

Identifying and Addressing Needs by Leveraging Resources

In 2002, USDA and EPA's Office of Pesticide Programs signed an agreement intended to leverage agency resources to improve coordination of federal oversight of GE crops that are engineered to tolerate herbicide treatments. Under the current regulatory framework, USDA regulates the herbicide-tolerant GE crop, while EPA regulates herbicides that are engineered into crops. In the 2002 agreement, USDA agreed to supply EPA with a list of herbicide-tolerant plants being field tested each year to ensure that EPA is aware of forthcoming products, and to provide EPA with a copy of petitions USDA receives from persons seeking nonregulated status for herbicide-tolerant crops. USDA also agreed to ask each applicant to submit a voluntary stewardship plan for the management of pest-resistance and weedy volunteer crops in herbicide-tolerant crop rotations and to consult with EPA on the viability of these stewardship plans. For its part, EPA agreed to supply USDA with current lists of herbicides registered for use on the crop in question and any readily available information about their efficacy.

However, we found that USDA and FDA could better leverage agency resources to address food safety issues for GE crops at the field trial stage. Specifically, FDA currently relies on GE crop developers to notify the agency that they are engaged in field trials of a plant with a novel trait or protein that might benefit from a voluntary early food safety evaluation. As the federal agency that reviews all applications for field trials of GE crops, USDA could alert FDA to field trials of such plants. At the same time, FDA could provide USDA with its evaluation of important food safety information, such as similarities between a new protein and known allergens and toxins and the overall stability of the protein, which USDA could use when making risk determinations for field trials of GE crops. Food safety concerns are one of several factors USDA takes into account when considering, for example, what types of permit conditions are needed for the environmental release of a GE crop, or whether activities associated with the crop should qualify for an exemption from the permit requirement. Currently, however, there are no formal mechanisms for coordinating the FDA early food safety evaluations with USDA's data on permits or notifications.

Establishing Compatible Policies, Procedures, and Other Means to Operate across Agency Boundaries

Although the specific procedures that the agencies use to regulate biotechnology vary according to each agency's legal authorities, the agencies hold interagency meetings to coordinate policies and share scientific information related to biotechnology across agency boundaries. There are currently two interagency groups that meet regularly to coordinate the federal government's oversight of agricultural biotechnology. One group is responsible for implementing the administration's policy on agricultural biotechnology and the other is a technical working group that provides agency officials involved in the day- to-day implementation of regulations with an opportunity to discuss emerging issues. These groups are as follows:

- *The Interagency Agricultural Biotechnology Working Group*. This working group, cochaired by OSTP and the National Economic

Council, was formed in 2001 to provide a forum for senior-level officials in relevant executive branch agencies—USDA, EPA, and FDA, the Office of Management and Budget, and the U.S. Trade Representative—to address agricultural biotechnology policy.[8] According to OSTP, the Biotechnology Working Group meets once a month or once every 2 months, as needed. Since its inception, the group has worked on several interagency initiatives, including coordinating negotiations between federal agencies to develop a coherent policy to address the low-level presence in food or feed of regulated GE crops. More recently, the group provided a forum for senior-level officials to discuss proposed regulatory revisions, such as the publication of USDA's *Draft Programmatic Environmental Impact Statement* and an EPA Advance Notice of Proposed Rulemaking, to address compliance issues for producers of GE pesticides. Also according to OSTP, when a major unauthorized release occurs, this group also provides a venue for officials to circulate information to ensure that the participating agencies are up to date on recent developments, and that the federal government's response is well-coordinated.

- *The Interagency Coordinated Framework Technical Working Group.* This working group was formed in 2003 to provide USDA, EPA, and FDA officials involved in the day-to-day implementation of regulations with an opportunity to meet monthly via conference call to discuss emerging issues. The group's past activities have included agency briefings on new GE products passing their respective approval or consultation processes, sharing information about upcoming rulemakings, and discussing lawsuits concerning the regulation of GE agricultural products.

Reinforcing Agency Accountability for Collaborative Efforts through Agency Plans and Reports

The agencies' strategic planning documents and performance reports do not specifically focus on the *Coordinated Framework* or the broad principles underlying the current regulatory system. However, these documents do address emerging issues related to biotechnology and recognize the need for interagency collaboration where appropriate. For example, USDA and FDA defined and measure their progress toward the shared goal of supporting international capacity building for agricultural biotechnology and promoting science-based oversight. In its strategic plan for 2005 through 2010, USDA established the goals of providing technical assistance and training to help countries adopt U.S. approaches to agricultural trade policy and helping foreign countries improve their regulatory structure for adopting biotechnology and agricultural biotechnology products. To measure its progress, USDA set a target of helping 15 countries make improvements to their trade policy and regulatory framework by 2010. Similarly, as part of its yearly report to stakeholders, FDA's Center for Food Safety and Applied Nutrition identified as priorities for 2007 its serving as the head of U.S. delegations and providing technical experts to two international task forces: (1) the Organisation for Economic Co-operation and Development's (OECD) Task Force on the Safety of Novel Foods and Feeds, which has worked to harmonize oversight of foods derived from biotechnology, and (2) the Codex Alimentarius Ad Hoc Intergovernmental Task Force on Foods Derived from Biotechnology, which has worked to develop a food safety assessment procedure for the low-level presence of regulated GE crops.[9] FDA also established the goal of providing technical assistance to the U.S. government on food biotechnology issues. EPA has been involved in similar initiatives, including participation on the previously mentioned task forces formed by OECD and Codex Alimentarius; however, EPA did not discuss these initiatives in the planning documents and reports that we reviewed.

Developing Mechanisms to Monitor, Evaluate, and Report on the Results

USDA and EPA have established mechanisms to help evaluate and report on matters related to the oversight of GE crops. Among the mechanisms established, USDA formed the USDA Advisory Committee on Biotechnology and 21st Century Agriculture in 2003 to provide information and advice to the Secretary of Agriculture on issues related to agricultural biotechnology. Since its inception, the committee has presented four consensus reports to the Secretary, including most recently a report on the issues that USDA should consider regarding the coexistence of GE, organic, and conventional crops. In addition, in response to USDA's requests, the National Research Council of the National Academy of Sciences has provided the agency with three science-based analyses of emerging issues in biotechnology, including GE crops.[10] EPA also has an advisory committee—the Federal Insecticide, Fungicide, and Rodenticide Act (FIFRA) Scientific Advisory Panel—that, while not specific to biotechnology, has provided recommendations and peer reviews related to EPA's oversight of GE pesticides on a number of occasions. The FIFRA

Scientific Advisory Panel, for example, played an important role in evaluating the health risks associated with the GE pesticide in StarLink corn.

However, several organizations have concluded that the agencies need better monitoring to detect unintended environmental or economic consequences and improve their risk analysis and management of marketed GE crops. In 2002, the National Research Council concluded that "screening of all crops with added genetic variation must be conducted over a number of years and locations because undesirable economic and ecological traits may only be produced under specific environmental conditions."[11] The council's report contained numerous recommendations regarding the monitoring of GE crops after they have been deregulated, including a recommendation that the federal government establish a long-term monitoring effort to assess potential environmental changes associated with the commercialization of GE crops, and that there be an open and deliberative process involving stakeholders to establish monitoring criteria. Similarly, in 2006, the National Science and Technology Council[12] cited the monitoring for ecosystem effects associated with the use of GE products as an area where the federal government could improve its risk assessments, noting that the ecological consequences are difficult to predict and that the variety of GE crops and organisms introduced in the environment is likely to grow.[13]

As an example of an unintended environmental consequence, EPA officials said that the widespread use of herbicide-tolerant GE crops could accelerate the development of herbicide-tolerant weeds. In this regard, weed scientists from Iowa State University and the University of Wisconsin said that federal support for mapping the occurrence of herbicide tolerance would be helpful. Another possible unintended consequence of the widespread use of crops containing the GE pesticide Bt is that Bt could lose its effectiveness against insect pests. As a condition of registering a Bt pesticide with EPA, registrants must require that users of the product follow certain insect-resistance management techniques, including planting "refuges" with non-Bt crops.[14] Registrants determine whether these requirements are met through surveys of farmers. However, some stakeholders with whom we spoke raised doubts about the effectiveness of having the registrant of a GE pesticide perform compliance monitoring activities.

Another concern stemming from the widespread use of GE crops is the economic impact they might have on farmers growing conventional or organic crops. For example, some growers of non-GE crops fear that seeds or pollen containing engineered traits from neighboring fields may commingle with their crops, thereby making those crops harder to sell to customers who prefer not to consume GE products. In this regard, in February 2007, the U.S. District Court for the District of Northern California ruled that USDA needed to conduct an environmental impact statement to analyze, among other things, the impact that deregulating a particular GE alfalfa might have on farmers growing organic or conventional alfalfa. In a 2008 report to the Secretary of Agriculture, USDA's Advisory Committee on Biotechnology and 21st Century Agriculture concluded that fostering coexistence between GE and non-GE crops is an important and worthwhile goal and acknowledged that the proximity of GE crops to conventional and organic crops sometimes causes commingling, preventing some retail consumers from finding products that are free of GE crops.[15] The committee recommended that the Secretary "take note" of several factors that can cause commingling, such as the failure to adequately contain regulated GE crops.

Despite these recommendations and observations from various sources, we found that USDA, EPA, and FDA do not have a mechanism to monitor, evaluate, and report on the impact of the commercialization of GE crops following the completion of the agencies' evaluation procedures. USDA, the agency with the most comprehensive authority regarding GE crops, has no systematic program of postmarket oversight. Once GE crops are deregulated, they are not subject to regulatory control under the Plant Protection Act, unless USDA finds them to be a plant pest or noxious weed on the basis of new data or analysis. EPA places conditions on the use of marketed GE pesticides, but its oversight is largely limited to the data it collects through the biotechnology developers that register the products. Without monitoring, undesirable agricultural and environmental problems could result from the unintended transfer of genetic material from deregulated GE crops to non-GE crops and other plants, and these problems could have significant financial implications.

Similarly, FDA generally does not monitor the use of GE crops in food or feed once they have been marketed. According to FDA officials, the agency does not routinely monitor the food supply for the presence of regulated GE crops because these crops may legally be present in food and feed, unless they contain an unapproved pesticide or food additive. In addition, as we have previously reported, monitoring the long-term health effects of GE food is generally neither necessary nor feasible, according to scientists and regulatory officials that we contacted.[16] In their view, such monitoring is unnecessary because there is no scientific evidence, or even a hypothesis, suggesting that long-term harm (such as increased cancer rates) results from these foods. Furthermore, there is consensus among these scientists and regulatory officials that technical challenges make long-term monitoring infeasible. Experts cited, for example, the technical inability to track the health effects of GE foods separately from those of their conventional counterparts. In addition, little is known about the longterm health effects of consuming most foods, meaning there is no baseline information against which to assess the health effects caused by GE foods. However, some stakeholders have expressed food safety concerns about the potential transfer of genetic material from food crops used to produce pharmaceutical and industrial compounds (see the following section on regulatory changes and other initiatives). While as of July 2008, the use of food crops to produce these compounds had not moved beyond limited field trials, in the future they may be produced on a larger scale for commercialization, increasing the potential for gene transfer to other crops and possible entry into the food and feed supply. This prospect suggests that some form of limited, directed monitoring of the food supply may be needed to ensure that these compounds are not present.

Officials Generally Did Not Cite Interagency Coordination as a Major Concern or Call for Revisions to the *Coordinated Framework*

In general, the officials from 22 stakeholder groups with whom we spoke did not indicate that interagency coordination was a major concern. Five officials told us that coordination among the agencies had improved over time. Nevertheless, some officials identified areas where interagency coordination could be improved. Most notably, five said that stronger central leadership, possibly residing in a high-ranking official, was needed to bring together the relevant agencies and to provide a unified government response to emerging issues and incidents as they occur. Two of these officials noted that such leadership existed in the past but has been inconsistent.

Similarly, the officials with whom we spoke generally did not identify changing the *Coordinated Framework* as a high priority. Of those that expressed an opinion, 10 officials told us that the framework has worked well and withstood the tests of time. On the other hand, four officials told us that the *Coordinated Framework* needed to be revised. Two of these four individuals, representing consumer advocacy organizations, said that using existing laws to govern biotechnology, as called for in the framework, was inadequate because agencies have had to “creatively interpret” or “bend over backward” to apply laws that do not specifically address biotechnology. They supported the creation of new laws specific to biotechnology. Meanwhile, six of the officials with whom we spoke did not have an opinion on the adequacy of the *Coordinated Framework*.

AGENCIES ARE CONSIDERING REGULATORY CHANGES AND OTHER INITIATIVES TO IMPROVE OVERSIGHT AND FURTHER LIMIT THE IMPACT OF POTENTIAL UNAUTHORIZED RELEASES

In recent years, USDA, EPA, and FDA have considered changes to which GE crops they regulate and how they will regulate them with the intention of improving oversight and reducing the impact of unintended releases of GE crops. In particular, in its July 2007 draft programmatic environmental impact statement (DEIS) and its October 2008 proposed rule, USDA is considering significant changes that could affect, among other things, which GE crops it regulates, requirements for pharmaceutical and industrial crops, and the agency's response to unauthorized releases. Proposals in the DEIS drew mixed views from stakeholders who submitted comments to the agency; the public comment period for the proposed rule was ongoing as we completed this report. Several of these proposed changes in the DEIS related to USDA's consideration of food safety and public health concerns, and some stakeholders have commented that USDA did not clearly state how it would coordinate human health assessments with FDA and EPA. USDA is also seeking funding to implement a voluntary quality management system designed to improve industry compliance with its field trial regulations. In addition, USDA has identified several operational lessons from its investigation of the LLRICE release that, if acted upon, could improve oversight. For its part, EPA has proposed amending several of its GE pesticide regulations, and stakeholders who submitted comments to the agency generally supported these proposals. Finally, FDA proposed in 2001 to make its voluntary premarket notification procedure mandatory; however, as of July 2008, the agency had not taken action to finalize the proposed rule, despite support from key stakeholders that we interviewed.

USDA Is Considering Significant Changes in How It Regulates GE crops

In July 2007, USDA published a DEIS outlining 10 issues related to biotechnology that may be the subject of future revisions to regulation. These 10 issues address such matters as which GE crops USDA should regulate, the permitting and notification process, the restrictions placed on GE crops that produce pharmaceutical compounds, and the agency's response to the low-level presence of regulated GE plant material. For each issue, USDA presented and assessed alternative regulatory approaches, including a no-action alternative and a preliminary preferred alternative. (See app. VIII for a list of the 10 issues and the alternatives that USDA assessed for each issue.)

The agency received 77 comments on the DEIS from stakeholders, such as individuals and organizations representing academia, the biotechnology industry, public interest groups, agricultural producers, and government agencies. USDA also received many comments from private citizens, 265 of which were submitted individually and another 23,379 that were in form letters forwarded by 2 public interest groups. We analyzed the comments from the 77 stakeholders as well as a random sample of 51 of the 265 comments submitted individually by private citizens.[17] Using the public comments and other considerations, USDA issued proposed amendments to its regulations in October 2008. According to USDA, differences between the proposed rule and the DEIS are primarily a matter of reorganizing and realigning some materials and their corresponding regulatory alternatives, using more descriptive terms in some criteria listed in the alternatives, and choosing between regulatory alternatives that fall within the analysis of the DEIS. Changes arising from this rulemaking process could represent the most extensive overhaul to the regulations since USDA originally implemented them in 1987. We selected 4 of the 10 issues addressed by the DEIS that we believe are particularly relevant to incidents of the unauthorized release of GE crops and analyzed the comments USDA received. While USDA has requested comments on the proposed rule, we were not able to review them for this report.[18]

Issue 1. Broadening Regulatory Scope to Include GE Crops Posing Noxious Weed Risk

This issue examines the question of which GE crops to regulate. Two alternatives USDA assessed in relation to this issue—including the one that the agency indicated was its preliminary preferred alternative—would expand USDA's oversight to all GE plants, not only those that pose a risk to plants.

These alternatives could also have allowed USDA the authority to consider the effect of GE crops on public health and the environment, rather than just the effect on other plants.

In our review of the stakeholder comments submitted to USDA, we found that 42 of the 44 stakeholders who indicated their preference supported expanding USDA's oversight. However, stakeholders had varied reactions to a key difference between the two alternatives—whether USDA should make regulatory decisions on an "event" or "trait" basis. Regulating by event would mean regulating each individual insertion or deletion of a gene or gene fragment from a cell. Regulating by trait would mean evaluating the characteristic (e.g., herbicide tolerance) manifested in the crop as a result of genetic engineering, and potentially making decisions for multiple events that exhibit the same trait. Some of those who favored regulating by event, which USDA indicated was its preliminary preferred alternative, believed it would be more protective. Some of those who favored regulating by trait indicated it would reduce the regulatory burden on developers. In its October 2008 rulemaking, USDA proposed to regulate GE plants on the basis of (1) known plant pest and noxious weed risks of the parent plants, (2) the traits of the GE plant, or (3) the possibility of unknown risks as a plant pest or noxious weed when insufficient information is available. Under the proposal, if adopted, USDA would encourage GE plant developers to consult the agency if they are uncertain whether a GE plant would be subject to regulation.

At the same time that USDA assessed the impact of expanding the reach of its regulatory oversight, it also assessed the impact of excluding certain classes of GE crops from regulatory oversight on the basis of risk. USDA included this exclusion—which 31 of the 37 stakeholders who expressed a preference supported—as part of its preliminary preferred alternative in the DEIS. Some stakeholders who commented on the DEIS and other observers have suggested that USDA could exclude from regulation plant pests from which disease-causing genes have been deleted. An example of a plant pest that is often used in genetic engineering is a bacterium known as *Agrobacterium tumefaciens* that can cause a plant disease known as crown-gall. When used in genetic engineering, its disease-causing genes are first removed. In the proposed rule, USDA stated that it anticipates that the range of GE plants subject to oversight will decrease as the agency reaches the conclusion that they do not pose increased or unfamiliar plant pest or noxious weed risks. The proposed rule also contains a procedure whereby the agency may approve petitions for conditional exemptions from permit requirements.

Issue 2. Use of Risk-Based Categories for New Products

Two of the alternatives under this issue that USDA considered in the DEIS would have expanded its current two-tier system of notifications and permits to further classify GE crops according to risk. Currently, USDA's policy allows for GE plants that meet specific eligibility criteria, such as cases in which the function of the introduced genetic material is known and does not result in plant disease, to be introduced under the notification process, while plants that do not meet the criteria must use the more stringent permit option. Under the two alternatives, USDA would have clarified and increased the number of tiers in which GE plants (and other GE organisms) could be placed. The DEIS proposed four tiers that would account for the potential of a GE plant to pose plant pest, noxious weed, or food safety risks. The tiers would impose different procedural requirements and permit conditions on GE crop developers. According to USDA, an expanded tier system would increase transparency and help focus agency resources on unfamiliar or high-risk crops. Almost all stakeholders (45 of 48) who expressed a preference preferred 1 of the alternatives that would expand the current 2-tiered system, with the remaining 3 stakeholders preferring that USDA abolish all categories and evaluate all field trial applications on a case-by-case basis. USDA's October 2008 proposed rule is consistent with the DEIS in that it would eliminate the notification procedure. USDA would continue to issue three types of permits for interstate movement, importation, and environmental release. The permits for environmental release of GE plants would be sorted into one of four categories on the basis of risk.

Issue 4. Regulation of Crops Producing Pharmaceutical and Industrial Compounds

USDA also assessed in its DEIS several alternatives for modifying its approach to issuing field trial permits for GE crops not intended for food or feed—namely, those engineered to produce pharmaceutical and industrial compounds. Currently, USDA imposes more stringent confinement and inspection requirements on these crops than it does for other types of GE crops. If unintentionally released into the food or feed supply, GE crops producing pharmaceutical and industrial compounds may pose risks to human health, trade, and the environment that are not posed by other types of GE crops, such as herbicide-resistant or insect- tolerant crops. USDA outlined a number of possible alternatives, such as prohibiting outdoor field tests of these crops or allowing only nonfood crops, such as tobacco, to be engineered to produce those compounds under the assumption that they would not be consumed inadvertently. Its preferred alternative was to continue to allow food and feed crops to be used for the production of pharmaceutical and industrial compounds, imposing confinement requirements as appropriate.

About half (27 of 52) of those stakeholders who expressed a preference, including all of the biotechnology developers and the majority of the academics and governmental organizations, preferred that USDA continue to allow food and feed crops (such as corn) to be used for the production of pharmaceutical and industrial compounds, but impose confinement requirements on the basis of the risk posed by the organism and consider food safety in setting permit conditions. However, 12 preferred that USDA prohibit outdoor field testing of GE crops engineered to produce pharmaceutical or industrial compounds if the type of crop used also has food or feed uses because of concerns that outdoor testing would increase the probability of those compounds spreading into the food or feed supply. Another 10 preferred that USDA prohibit outdoor field testing of any GE crop that produces these compounds. The remaining 3 advocated prohibiting the use of food and feed crops, regardless of whether the crop is grown in an outdoor field test or in a contained facility. Of all of the issues discussed in the DEIS, this is the one that most concerned private citizens. The Union of Concerned Scientists and the Center for Food Safety forwarded almost 23,400 comments from private citizens urging USDA to ban the outdoor production of pharmaceutical and industrial compounds in food crops. In its proposed rule, USDA concluded that its proposed permitting procedures and the use of stringent permit conditions can effectively minimize the risks that might be associated with the environmental release of GE plants that produce pharmaceutical or industrial compounds, including GE plants that are normally food crops.

Issue 7. Allowance for Low- Level Presence of Regulated GE Material in Crops, Food, Feed, or Seed

Under this issue, USDA's DEIS evaluated alternatives that would establish criteria for determining that an unauthorized release of a low level of regulated GE crops outside of a field trial site is "nonactionable"—that is, determining when a GE crop poses a low risk to health or the environment. Currently, the agency's policy is to respond to incidents of low-level presence on a case-by-case basis, assessing the specific health and environmental risks posed by the regulated material and taking remedial action only when necessary. In its DEIS, USDA proposed specific criteria under which the agency would not take remedial action in response to unauthorized releases that pose minimal risk, contending that these criteria would reassure the public and other countries of the safety of any regulated GE crops detected at low levels in commercial plants or seeds. The majority of stakeholders (34 of 46) who expressed a preference supported establishing criteria for determining when a release is nonactionable. A number of academic stakeholders attributed the market disruptions that followed unauthorized releases to a perception of risk created by the current regulations, which treat all releases alike. Some stakeholders also noted that tolerances have been developed to allow for low levels of contaminants, such as pesticides or insect parts, in the food supply, and that USDA should be able to develop similar tolerances for GE crops that pose no known risk to human health. In addition, some stakeholders supported the relaxation of confinement standards in some instances, arguing that the low-level presence of genes moving from a GE plant to a non-GE plant should not, in itself, be a concern because gene flow is a pervasive and naturally occurring process.

However, other stakeholders (7 of 46), namely public interest groups and agricultural producers, supported the most stringent of the proposed alternatives, which would impose confinement requirements on all GE crops comparable to those now imposed on pharmaceutical and industrial crops and would consider all low-level presence to be actionable. These stakeholders argued that trace amounts of regulated material could jeopardize organic agriculture, particularly in export markets that have tighter standards, and that USDA does not have sufficient scientific data on the long-term effects of GE crops to make the determination that low levels are safe. The private citizens whose comments we analyzed and who expressed a preference also preferred this alternative.

USDA's October 2008 proposed amendment to its regulations is generally consistent with the preliminary preferred alternative in the DEIS. USDA proposes to investigate each incident of low level presence individually before making a decision on what, if any, remedial action is needed. USDA would use specific criteria enumerated in the proposed regulations to rate the risk involved in the incident. However, those criteria would not fully determine the agency's response; USDA would evaluate other relevant information and order remedial action if it appears necessary.

Stakeholders Raised Concerns That USDA Did Not Clearly State How It Would Coordinate Human Health Assessments with EPA and FDA

Four of the 10 issues described in USDA's DEIS referred to the agency's consideration of food safety and public health concerns associated with GE crops, and some stakeholders thought it was unclear whether it would be USDA's, EPA's, or FDA's responsibility to perform the necessary evaluations. These 4 issues are as follows:

- In issue 1, regarding the broadening of USDA's scope of oversight of GE crops, as we have previously discussed, USDA's preliminary preferred alternative in its DEIS as well as its proposed regulatory amendment would use the agency's authority to consider the effect that GE crops could have on public health. To date, USDA has regulated GE crops on the basis of their risk as a plant pest—it takes into consideration human health data when responding to petitions to deregulate GE crops to meet NEPA requirements, but FDA has primary responsibility for food safety. Under its proposed rule, USDA would use its authority under the Plant Protection Act to regulate GE crops as potential noxious weeds, which would enable it to regulate crops on the basis of their effect on public health. For example, it could consider public health in setting the conditions for field trials of GE crops and could require that all food safety issues be resolved prior to deregulation. However, the agency did not provide specific details in either the DEIS or the proposed rule on how it intends to evaluate human health effects or determine when food safety issues have been resolved.
- In issue 2, USDA's preliminary preferred alternative in the DEIS, as well as its proposed regulatory amendment, would include human health as a criterion for determining which category a GE crop would fall into under the proposed risk-based system. For example, according to the DEIS, to qualify for the lowest risk tier, a GE food would need an EPA-issued pesticide tolerance or an alternative evaluation of its toxicity and allergenicity. In general, under its preferred alternative, USDA would consider the toxicity and allergenicity of GE crops when imposing confinement requirements on field test sites. However, USDA did not specify which agency would evaluate toxicity or allergenicity.

- In issue 4, USDA's DEIS described several alternatives for regulating crops engineered to produce pharmaceutical and industrial compounds, including one alternative that would use evaluations of food safety to determine the appropriate confinement measures, and another alternative that would require that food safety concerns be addressed prior to the use of a food or feed crop for the production of such compounds. However, USDA did not describe the role that FDA or EPA, the agencies that have primary responsibility for regulating pharmaceutical and industrial compounds, respectively, would have in providing health assessments of GE crops used for these purposes.
- In issue 7, USDA indicated in the DEIS that one potential criterion for determining whether the low-level presence of a regulated GE crop in the food supply or the environment is nonactionable would be if food safety issues have been adequately addressed. However, in the DEIS, USDA did not indicate how it would use food safety assessments from other agencies, such as FDA or EPA, in deciding whether a low-level presence is nonactionable. In its October 2008 rulemaking, USDA proposed that for food and feed crops, one of the following three conditions must be true for the agency to determine that a low level presence is nonactionable: (1) EPA has established a tolerance or an exemption from tolerance for any GE pesticide expressed by the GE plant, (2) key food safety issues of the new protein or other substance have been addressed, or (3) no new protein or substance is produced.

A range of stakeholders, including academics, state officials, and public interest groups, commenting on the DEIS expressed concern that if USDA decides to evaluate the public health consequences of new GE crops, its oversight responsibilities would overlap with those of EPA and FDA.

Several stakeholders encouraged USDA to coordinate its regulatory activities with those of EPA and FDA when addressing human health concerns. USDA acknowledged that addressing all of the food safety issues discussed would likely increase these agencies' workload. USDA's DEIS did not describe how it would incorporate other agencies' programs, such as FDA's early food safety evaluations of novel proteins, into its oversight. In addition, FDA's early food safety evaluations do not apply to crops intended exclusively for the production of pharmaceutical or industrial compounds. FDA officials said they had no plans to perform such evaluations in the future.

In its October 2008 proposed rule, USDA acknowledged FDA's authority in the food safety area, but also emphasized the need for mutual agency support. USDA stated that it would evaluate permit applications for new GE organisms, including plants, to determine if they could present risks to the public health. If so, USDA would contact FDA. The decision to regulate food and feed from the GE organism would be FDA's. USDA also stated that it would take into account existing food safety evaluations when evaluating GE organisms.

USDA Seeks to Establish a Voluntary Biotechnology Quality Management System to Help Improve Industry Compliance with Field Trial Regulations

USDA also is seeking $4.0 million in additional funding for fiscal year 2009 to establish a quality management system to improve developers' compliance with field trial regulations.[19] USDA has concluded that there is a lack of quality management systems among GE plant developers, and, in September 2007, the agency announced that it would establish a voluntary program called the Biotechnology Quality Management System (BQMS) to help universities, small businesses, and large companies develop policies and practices that will enable them to proactively address potential compliance problems before they materialize. Participants would identify vulnerabilities in their processes, develop quality control measures to minimize the risk of unauthorized releases, and demonstrate—through recordkeeping and a documented management system—their ability to manage the safe introduction of GE crops into the environment. In addition, USDA would (1) work with permit holders to ensure that quality management plans are developed and in place, (2) develop standardized quality assurance and best practices guidance documents, and (3) provide outreach to the regulated community. USDA's Agricultural Marketing Service would manage the audit component of the program and accredit third-party auditors. However, BQMS would not replace USDA's existing regulatory compliance and inspection process.

An issue raised by several members of USDA's advisory panel on biotechnology was whether BQMS's benefits would justify its likely costs to the government and regulated community. One particular concern was that the program, while called voluntary, would become an expensive *de facto* mandatory program for developers with limited resources, such, as universities, if agencies used participation in the program as a criterion for awarding federal funding for GE research. Another concern was whether there would be adequate incentives to encourage participation. However, in its 2008 report on coexistence, the advisory committee also concluded that programs like BQMS may help address factors that inhibit coexistence among different agricultural production systems, including the production of GE, conventional, and organic crops.

USDA Has Identified Lessons Learned from the LLRICE Incidents That Could Improve Oversight

In October 2007, USDA issued a compilation of proposed changes intended to enhance its oversight of GE crops on the basis of lessons learned from its investigation of the LLRICE incidents and its 20 years of experience in GE crop regulation. The lessons learned related to a range of issues, including inadequate record keeping by permit and notification holders, delays in obtaining representative samples of GE seed, developers' lack of corrective action plans, incomplete access to agreements made among GE crop developers and entities they have contracted with to conduct field trials, and the sufficiency of isolation distances between field trial sites and other crops. USDA also noted that it lacked the authority to subpoena anything other than documents—for example, the agency could not subpoena seeds or plant parts. The recently enacted 2008 Farm Bill contains language directing the Secretary of Agriculture to take action on the lessons learned within 18 months. According to USDA, its October 2008 proposed rulemaking would address many of the Farm Bill requirements, particularly as it relates to recordkeeping and reporting. The 2008 Farm Bill also expanded USDA's subpoena authority to cover "tangible things that constitute or contain evidence."

EPA Has Proposed Amending Its Regulations for GE Pesticides

EPA is considering amending its regulations governing GE pesticides. In April 2007, EPA proposed two related rules intended to create a more risk-based system for regulating a certain type of GE pesticide known as a plant virus coat protein (PVCP).[20] In the first rule, EPA proposed to exempt from regulation PVCPs that present minimal risk to human health or the environment. In the second rule, the agency proposed to exempt from regulation the residues produced by GE pesticides that are based on viral coat proteins. Under these rules, developers would be able to self- determine whether a new PVCP-based GE pesticide is exempt from EPA's pesticide registration requirements and the requirement of a pesticide tolerance on the basis of specific risk-based criteria established by EPA. Stakeholders commenting on EPA's proposed rules had mixed views on the scope of the agency's proposals. In addition to those that supported the exemptions proposed by EPA, there were stakeholders from scientific associations that favored extending the exemption to plant virus genes other than virus coat proteins, as well as those that favored limiting the exemption to certain types of PVCP GE pesticides. On the other hand, some comments from food industry and safety organizations expressed concern about EPA's proposed exemptions, citing scientific uncertainty.

Also in April 2007, EPA published an advance notice of proposed rulemaking describing possible revisions that would help EPA account for the differences between GE pesticides and conventional pesticides and help ensure that developers of GE pesticides comply with necessary requirements. According to EPA, current regulations for agricultural pesticides were written before GE pesticides were defined, and may not adequately address the distinction. As such, they may not apply to the unique characteristics of GE pesticides produced in a GE crop on a farm. Specifically, EPA is considering amending regulations governing the (1) registration of GE pesticide production facilities, (2) reporting and record-keeping requirements, (3) issuance of experimental use permits, and (4) requirements for labeling.

Most stakeholders supported these proposals to distinguish between GE pesticides and other pesticides. For example, most stakeholders favored modifying the current definitions in FIFRA relating to GE pesticides, either by excluding farmers and seed processors from the current definition of pesticide "producer" and "establishment" or by including other parties, such as companies that license a GE pesticide for inclusion in plants, facilities that produce seeds containing GE pesticides, and any laboratory or greenhouse where a pesticide is engineered into a plant. However, some stakeholders had varied views about changes to labeling requirements for GE pesticides. Under current practices, according to EPA officials, seed labels do not need to identify that the seed contains a registered pesticide that might have certain use restrictions. Instead, EPA requires as a condition to registration that registrants ensure that growers comply with any planting restrictions associated with the seed. For example, growers are expected to sign a contract with the registrant of the pesticide agreeing to certain planting restrictions as well as routine "compliance assurance visits." While some (4) felt this system was adequate, others (3) thought a legally enforceable label would help promote growers' compliance with planting restrictions.

FDA Proposed in 2001 to Make Premarket Notification Mandatory

In 2001, FDA proposed a rule that would require companies to notify the agency before marketing GE crops as food or feed products to complement its voluntary consultations. Among the reasons that FDA cited for proposing this change were concerns expressed by consumers and public interest groups about the limits to the transparency and the voluntary nature of the consultation process and the potential of genetic engineering to create more complex safety issues. Many stakeholders with whom we spoke were in favor of this proposal. For example, a representative from the Grocery Manufacturers Association/Food Products Association said that food safety assessments should be mandatory and done early enough so that the public could be assured of product safety if regulated articles were unintentionally released into the food supply. Similarly, a representative from the rice industry also said that food assessments should be mandatory and that if a premarket notification had been done for LLRICE, it would have reduced the economic impacts of unauthorized releases. However, as of July 2008, FDA had not taken action to finalize this proposed rule, and FDA officials told us that such a rule no longer may be needed because the voluntary consultation process is working well and fully protects the public health.

CONCLUSIONS

After two decades of experience with field trials, it is widely acknowledged that unauthorized releases of regulated material from field trial sites are likely to occur in the future, and, accordingly, releases are one area of the *Coordinated Framework* that has been reviewed and modernized in recent years. While the OSTP's 2002 policy document outlines important first steps for agencies to take to address the likelihood of the low-level presence of regulated genetic material in the environment or food supply and to mitigate any potential economic, environmental, or human health consequences, there are two areas where the agencies could improve their implementation of these proposals, as follows.

- First, FDA has yet to make publicly available, as was initially intended, the results of its early food safety evaluations of novel proteins engineered into plants. In the absence of timely information about the actual risks to human health and the environment presented by a GE crop in the field trial stage, FDA may be missing an opportunity to mitigate the impacts of unauthorized releases, enhance the agency's credibility, and improve public confidence.
- Second, USDA and FDA have not taken steps to fully leverage their resources to address food safety issues for certain GE crops at the field trial stage. While the agencies have acted to implement the proposals in OSTP's 2002 policy document to address field trials of GE crops, a lack of coordination of key information among the agencies may prevent them from making the most effective use of their resources. Specifically, the agencies do not have a formal mechanism for sharing information that could enhance their oversight of GE crops in the field trial stage that contain new proteins and that, if released into the food supply, could cause health concerns. FDA currently relies on crop developers to voluntarily notify the agency that they are engaged in field trials of a plant that might benefit from an early food safety evaluation. Because USDA, the federal agency that reviews all applications for field trials of GE crops, does not have a formal mechanism to alert FDA to field trials of such plants, FDA is less likely to be aware of developers' activities and to encourage them to participate in an evaluation. At the same time, without a formal mechanism for sharing the results of FDA's evaluations, USDA may lack important food safety information that it could use when making risk determinations for field trials of GE crops and when setting confinement and remediation measures.

To date, government oversight of GE crops has largely focused on assessing and preventing risks posed by GE crops in the testing phase, assuming that after GE crops enter commercial production, the need to oversee them diminishes. However, as the volume and variety of GE crops being grown increases, many stakeholders, including the National Research Council and the National Science and Technology Council, are becoming concerned that widespread use of GE crops can have unintended consequences that should be monitored. The consequences could include negative effects on the environment, non-GE segments of agriculture, or food safety. Among the practices we have identified as important to enhancing collaboration among agencies is developing mechanisms to monitor, evaluate, and report on the results of agency decisions. Such mechanisms should be applied to decisions that lead to the commercial use of GE crops. However, such a monitoring program should be based on risk. Not all GE crops that are marketed may warrant monitoring, and the duration of monitoring may not need to be indefinite.

Recommendations for Executive Action

To improve transparency and mitigate the impact of an unauthorized release into the food or feed supply of a regulated GE plant that has completed an early food safety evaluation, we recommend that the FDA Commissioner fulfill the agency's commitment to post the results of completed early food safety evaluations on its Web site and add the results of future evaluations within 120 days of receiving the submission from the plant developer.

To reduce the risk and impact of unauthorized releases, we recommend that the Secretary of Agriculture and the FDA Commissioner develop a formal agreement to share information concerning GE crops with novel genetic traits that, if unintentionally released into the food or feed supply, present or are likely to present public health concerns and, as a result, also could have negative financial consequences for the food and agriculture industry. With information from USDA about permits or notifications for field trials of such GE crops, FDA could identify which GE crops might benefit from an early food safety evaluation and encourage the developers of those crops to participate in evaluations. With assistance from FDA, USDA could make meaningful and transparent use of the health evaluation data available through FDA's early food safety evaluations in its risk assessment of GE crops.

To help ensure that unintended consequences arising from the marketing of GE crops are detected and minimized, we recommend that the Secretary of Agriculture, the EPA Administrator, and the FDA Commissioner develop a coordinated strategy for monitoring marketed GE crops and use the results to inform their oversight of these crops. Such a strategy should adopt a risk-based approach to identify the types of marketed GE crops that warrant monitoring, such as those with the greatest potential for affecting the environment or non-GE segments of agriculture, or those that might threaten food safety through the unintentional introduction of pharmaceutical or industrial compounds into the food supply. The strategy should also identify criteria for determining when monitoring is no longer needed. In developing a strategy, the agencies should draw upon the analysis and conclusions of the National Research Council and the National Science and Technology Council.

AGENCY COMMENTS AND OUR EVALUATION

We provided a draft of this chapterto USDA, EPA, and the Department of Health and Human Services (FDA) for review and comment. USDA and FDA provided written comments; these comments are reproduced in appendixes II and III, respectively. EPA provided its comments orally. The agencies generally agreed with the report's findings. FDA and EPA also provided technical comments that we have incorporated as appropriate. In addition, we provided a draft of this chapterto the Office of the United States Trade Representative for informal review and comment. This Office responded that it had no comments on the report.

Concerning our first recommendation, FDA said that it intends to make every effort to fulfill its commitment to post the results of completed early food safety evaluations on its Web site and add the results of future evaluations within 120 days of receiving the submission from the plant developer. However, FDA also said that activities of greater public health priority have been the focus of its limited resources. While acknowledging these priority and resource considerations, we continue to believe that implementing this recommendation would be a relatively low-cost way to increase public transparency and trust and mitigate the impact of the unintended release of GE crops subject to early food safety evaluations.

Regarding our second recommendation, USDA and FDA agreed, in part, that developing a formal agreement could enhance the sharing of information concerning GE crops with novel genetic traits that, if unintentionally released into the food or feed supply, could cause health concerns and have negative financial consequences. For example, USDA stated that information obtained from FDA under this agreement could assist USDA in its decisions on confinement conditions and deregulation of certain GE organisms. FDA also said that it would be useful to explore possible mechanisms for sharing information with USDA. However, the agencies said they should focus their resources on issues that present or are likely to present public health concerns rather than issues that pose only "perceived" concerns. In addition, regarding the financial consequences of unintended releases, FDA said this possibility falls outside the scope of its authority to protect and promote the public health. However, we note that USDA, which bears some responsibility for promoting and expanding agricultural markets, may be concerned with these consequences. Because sharing such information would be beneficial, we retained the reference to the financial consequences of unintended releases in the recommendation. As we reported, the known unintended releases of GE crops into the food or feed supply apparently have not caused health effects, but several led to financial losses. Nonetheless, we modified this recommendation to remove the reference to "perceived health concerns" and instead emphasize that the agreement would cover GE crops that present or are likely to present public health concerns.

USDA, EPA, and FDA agreed, in part, with the third recommendation that they develop a coordinated strategy to monitor marketed GE crops for unintended consequences. USDA stated that it supports having discussions with EPA and FDA regarding monitoring strategies for marketed deregulated GE crops. While USDA agreed that monitoring a partially deregulated GE crop might be appropriate where a potential plant pest risk is identified, USDA said its current regulations limit it to monitoring only regulated crops, and only for plant pest risks. We note that USDA maintains authority under the Plant Protection Act to regulate GE crops that it previously deregulated if it obtains new information indicating the crop is a plant pest. We also note that USDA has authority under the Plant Protection Act to regulate GE crops as noxious weeds, if warranted. Finally, in light of known unauthorized releases that led to financial losses, we believe that USDA should contribute to monitoring for other unintended consequences, such as economic impacts on other agriculture sectors, such as organic crops, that may become contaminated by GE crops. Also regarding monitoring strategies for marketed deregulated crops, EPA said that it intends to discuss such coordination issues with USDA and FDA to be better prepared in case a situation should arise in the future that warrants monitoring and is willing to continue working with the other agencies to determine whether additional monitoring mechanisms are worthy of consideration, how such monitoring would be conducted, and what resources would be required. However, EPA opined that GE crops that produce pesticides do not require any post- market monitoring beyond what is currently in place. For example, EPA noted that companies are required by FIFRA to report any adverse effects associated with GE pesticides and, in some cases, EPA has required companies holding registrations for GE pesticides to conduct studies on their effects. While acknowledging these monitoring mechanisms already in use, we still believe the agencies need a coordinated strategy for monitoring marketed GE crops that could include, in part, these mechanisms. FDA said that post-market monitoring of foods derived from GE crops is not necessary, but that it would consider risk-based monitoring should marketed GE crops intended for food or feed warrant such scrutiny in the future. FDA also indicated that it plans to discuss coordination issues with the other agencies to be better prepared should such a situation arise. In making this recommendation, our concern, in part, was the potential for GE crops producing pharmaceutical or industrial substances to be inadvertently present in the food or feed supply. In that regard, FDA opined that random sampling to detect pharmaceutical or industrial substances would present significant technical challenges and greatly affect resources and would be less effective than USDA's current system of strict permit conditions and inspections targeted to GE crops used to produce these substances. However, given that in the United States (1) GE crop varieties are grown extensively, (2) most processed foods contain ingredients from GE crops, (3) it is inherently difficult to prevent the spread of plant genetic material in the environment, (4) there may be an increasing use of GE crops to produce an even wider array of pharmaceutical and industrial compounds in the future, and (5) genetic modifications are becoming increasingly complex in response to pressures to increase yields for food and biofuel, we stand by our recommendation that the agencies should develop a coordinated strategy for risk-based monitoring of marketed GE crops.

As agreed with your offices, unless you publicly announce the contents of this chapterearlier, we will plan no further distribution until 30 days from the report date. At that time, we will send copies of this chapterto the appropriate congressional committees; the Secretary of Agriculture; the Administrator of EPA; the Secretary of Health and Human Services; the Commissioner of FDA; the Director, Office of Management and Budget; and other interested parties.

Lisa Shames

Director, Natural Resources and Environment

APPENDIX I. OBJECTIVES, SCOPE, AND METHODOLOGY

At the request of the Chairman and Ranking Member, Senate Committee on Agriculture, Nutrition, and Forestry, we evaluated federal oversight of genetically engineered (GE) crops. Specifically, our objectives were to examine (1) unauthorized releases of GE crops into food, feed, or the environment; (2) the degree of coordination among the federal agencies that regulate GE crops under the 1986 *Coordinated Framework for Regulation of Biotechnology (Coordinated Framework)*; and (3) additional actions the agencies have proposed to improve the oversight of GE crops and reduce the potential for unauthorized releases. The focus of our work was on the federal regulation of GE crops. We did not assess regulation of GE animals or other nonplant organisms. In addition, we did not assess U.S. efforts to reduce barriers to international trade in GE agricultural commodities.

In general, to achieve our objectives, we interviewed officials or obtained documentation from relevant federal agencies, including the U.S. Department of Agriculture (USDA); Environmental Protection Agency (EPA); Food and Drug Administration (FDA), and Office of Science and Technology Policy (OSTP) which is within the Executive Office of the President, as well as from agriculture and food industry and consumer organizations. Industry organizations included the American Farm Bureau Federation, American Seed Trade Association, American Soybean Association, Association of Official Seed Certifying Agencies, Grocery Manufacturers' Association/Food Products Association, National Association of Wheat Growers, National Corn Growers Association, National Grain and Feed Association, North American Export Grain Association, Organic Trade Association, U.S. Soybean Export Council, and USA Rice Federation. Consumer organizations included the Center for Food Safety, Center for Science in the Public Interest, Consumers Union, and Union of Concerned Scientists. In addition, we interviewed officials or obtained documentation from the Biotechnology Industry Organization; biotechnology companies, such as Arborgen and Monsanto; academics involved in genetic engineering research; the National Research Council; and the Pew Initiative on Food and Biotechnology.

More specifically, to examine unauthorized releases of GE crops into food, feed, or the environment, we reviewed government documents, academic literature, and media accounts related to known incidents of releases. We also discussed these incidents and their potential environmental, financial, health, and trade implications with industry, consumer, and academic officials. Furthermore, to examine the federal government's role in preventing unauthorized releases and mitigating their impact, we reviewed relevant laws and regulations and discussed their implementation with USDA, EPA, and FDA officials. At USDA, we also reviewed data on field trial permits and inspections done during fiscal years 2005 through 2007, and data on suspected violations and enforcement actions taken during fiscal years 2003 through 2007. In addition, we reviewed case files on potential incidents of unauthorized releases reported during fiscal years 2003 through 2007 that were referred for investigation. We also reviewed a random sample of other case files that were resolved without an investigation during this period. In addition, at EPA, we reviewed data on field trial permits issued from fiscal year 1997 through May 2008, and documentation on the four enforcement actions taken from fiscal year 1996 through August 2008. Since EPA had delegated enforcement authority, including the responsibility for doing field trial inspections to all 50 states except Wyoming, the agency was unable to provide us with summary data on the number of completed inspections involving GE pesticides.

To determine the degree of coordination among agencies that regulate GE crops, we reviewed the *Coordinated Framework's* guidance for interagency coordination. We then discussed with agency officials their implementation of this guidance and reviewed documents that they provided, such as interagency memorandums of understanding and agendas or minutes for interagency meetings. We also considered the views of nongovernmental organizations regarding the adequacy of this coordination, including those published by the National Research Council and the Pew Initiative on Food and Biotechnology. In addition, for criteria, we applied selected practices previously identified by GAO for enhancing and sustaining interagency collaboration.[21] These practices include defining and articulating a common outcome; agreeing on roles and responsibilities; establishing mutually reinforcing or joint strategies; identifying and addressing needs by leveraging resources; establishing compatible policies, procedures, and other means to operate across agency boundaries; developing mechanisms to monitor, evaluate, and report on results; and reinforcing agency accountability for collaborative efforts through agency plans and reports. We did not address an eighth practice—that is, reinforcing individual accountability for collaborative efforts through performance management systems—because doing so was beyond the scope of our work.

Related to the coordination issue and other aspects of the *Coordinated Framework*, we also compared the guidance contained in the framework and in related policy statements subsequently issued by OSTP with regulations and proposed rules promulgated by USDA, EPA, and FDA since 1986. In addition, we discussed the framework's relevance with industry, consumer, and academic officials. Although the framework is a broad policy document addressing all aspects of biotechnology, our analysis was limited to those sections that pertain specifically to the regulation of GE crops.

To determine the additional actions proposed by the agencies to improve oversight of GE crops and reduce the potential for unauthorized releases, we reviewed relevant proposed rules published in the *Federal Register*. These proposed rules included the following:

- USDA, *Proposed Rule: Importation, Interstate Movement, and Release Into the Environment of Certain Genetically Engineered Organisms, 73 Fed. Reg.* 60,008 (Oct. 9, 2008).
- USDA, *Introduction of Genetically Engineered Organisms, Draft Programmatic Environmental Impact Statement,* (July 17, 2007).
- EPA, *Proposed Rule: Exemption from the Requirement of a Tolerance under the Federal Food, Drug, and Cosmetic Act for Residues of Plant Virus Coat Proteins that are Part of a Plant-Incorporated Protectant (PVC-Proteins), 72 Fed. Reg.* 19,640 (Apr. 18, 2007).
- EPA, *Proposed Rule: Exemption Under the Federal Insecticide, Fungicide, and Rodenticide Act for Certain Plant-Incorporated Protectants Derived From Plant Viral Coat Protein (PVCP-PIPs) Gene(s) Supplemental Proposal, 72 Fed. Reg.* 19,590 (Apr. 18, 2007).
- EPA, *Advance Notice of Proposed Rulemaking: Plant-Incorporated Protectants; Potential Revisions to Current Production Regulations, 72 Fed. Reg.* 16,312 (Apr. 4, 2007).
- FDA, *Proposed Rule: Premarket Notice Concerning Bioengineered Foods, 66 Fed. Reg.* 4,706 (Jan. 18, 2001).

USDA issued its October 9, 2008 proposed rule after we had sent our draft report to the agencies for review and comment. We modified our draft to reflect the publication of the proposed rule, and added brief descriptions of some aspects of it. However, we were not able to thoroughly analyze the proposed rule or interview agency or other stakeholder officials about its contents.

We also reviewed public comments submitted with respect to each of these proposed rules except USDA's October 2008 proposed rule. (The deadline for commenting on that proposed rule is November 24, 2008.) In general, these comments are posted to each rule's official electronic docket found at *Regulations.gov*. To summarize the comments on USDA's draft programmatic environmental impact statement (DEIS), we divided the respondents into seven constituent categories: academics, agricultural producers, biotechnology developers, consumer and public interest groups, food industry representatives, government officials, and unaffiliated private citizens. A GAO analyst then coded the responses from the first six constituent categories on the basis of stakeholders' explicit or implied preference for various alternatives discussed in the draft statement. The coding scheme included a means of indicating when a stakeholder's preference was not apparent on the basis of the written comments. To ensure that decisions about how to code the comments were reliable, a second GAO analyst also reviewed the comments. We used the same technique to code a random sample of 51 of the 265 comments submitted individually by unaffiliated private citizens.

To summarize the views of stakeholders who commented on EPA's proposed rules, we coded all stakeholders' comments on the basis of their general response to the rules as well as their specific responses to relevant issues identified by EPA. Because of the limited number of responses—generally about 12—posted in each docket, we did not stratify respondents into different categories. Regarding FDA's proposed rule, we could not easily stratify and summarize the associated comments, which, according to FDA, numbered over 124,000. Specifically, as of August 2008, FDA had not entered these comments into an electronic docket that we needed to perform this analysis. Instead, we reviewed a limited, judgmental sample of these comments to gain a general understanding of the issues that stakeholders raised.

Furthermore, to determine additional actions proposed by USDA, we interviewed agency officials and reviewed documentation they provided related to two initiatives—that is, USDA's (1) proposal for a voluntary Biotechnology Quality Management System (BQMS) and (2) summary of lessons learned from its investigation of the unauthorized release of a GE rice variety, LibertyLink Rice (LLRICE), and other similar incidents. BQMS, which USDA plans to fully implement in fiscal year 2009, provides guidance to GE crop developers for analyzing their field trial operations to identify possible problems and mitigation measures that could reduce the potential for an unauthorized release. Also, we attended meetings in November 2007 and March 2008 of USDA's Advisory Committee on Biotechnology and 21st Century Agriculture at which the BQMS proposal was discussed.

We conducted this performance audit from July 2007 to November 2008 in accordance with generally accepted government auditing standards. These standards require that we plan and perform our audit to obtain sufficient, appropriate evidence to provide a reasonable basis for our findings and conclusions based on our audit objectives. We believe that the evidence obtained provides this reasonable basis.

APPENDIX II. COMMENTS FROM THE U.S. DEPARTMENT OF AGRICULTURE

Note: GAO comments supplementing those in the report text appear at the end of this appendix.

USDA

United States Department of Agriculture

Office of the Secretary
Washington, D.C. 20250

OCT 31 2008

2008 OCT 31 AM 10:45

US GAO

Mr. Jim Jones, Assistant Director
United States Government Accountability Office
441 G Street, NW
Washington, DC 20548

Dear Mr. Jones:

The United States Department of Agriculture (USDA) appreciates the opportunity to review and provide comments on the GAO Draft Report, "Genetically Engineered Crops: Agencies Are Proposing Changes to Improve Oversight, but Could Take Additional Steps to Enhance Coordination and Monitoring," (GAO 09-60). We have addressed the Recommendations for Executive Action that pertain to USDA.

GAO Recommendation

To reduce the risk and impact of unauthorized releases, we recommend that the Secretary of Agriculture and the Commissioner of FDA develop a formal agreement to share information concerning GE crops with novel genetic traits that, if unintentionally released into the food or feed supply, could cause real or perceived health concerns and have negative financial consequences for the food and agriculture industry. With information from USDA about permits or notifications for field trials of such GE crops, FDA could identify which GE crops might benefit from an early food safety assessment and encourage the developers of those crops to participate in assessments. With assistance from FDA, USDA could make meaningful and transparent use of the health assessment data available through FDA's early food safety assessments in its risk assessment of GE crops.

USDA Response

USDA agrees, in large part, with this Recommendation. Under the Coordinated Framework, USDA and the U.S. Food and Drug Administration (FDA) have worked effectively to share information on this and related issues. However, development of a formal agreement between USDA and FDA could enhance relevant information sharing.

USDA will work with FDA to explore the development of an agreement for sharing information to help ensure that each agency has relevant information concerning regulated GE crops. USDA recently published a proposed rule to revise existing regulations regarding the importation, interstate movement and environmental release of certain genetically engineered (GE) organisms under the Plant Protection Act (PPA) of

regulations regarding the importation, interstate movement and environmental release of certain genetically engineered (GE) organisms under the Plant Protection Act (PPA) of 2000. During the comment period, USDA may receive additional suggestions related to enhancing information sharing with FDA. Information that is shared by FDA under an agreement may assist USDA in its decisions on confinement conditions and deregulation of certain GE organisms. Nonetheless, USDA's regulatory system is science-based, and therefore its evaluations and decision making are, and will remain, limited to risks involving actual physical harm and will not consider perceived, but scientifically unfounded, risks.

GAO Recommendation

To help ensure that unintended consequences arising from the marketing of GE crops are detected and minimized, we recommend that the Secretary of Agriculture, the Administrator of EPA, and the Commissioner of FDA develop a coordinated strategy for monitoring marketed GE crops and use the results to inform their oversight of these crops. Such a strategy should adopt a risk-based approach to identify the types of marketed GE crops that warrant monitoring, such as those with the greatest potential for affecting the environment or non-GE segments of agriculture, or those that might threaten food safety through the unintentional introduction of pharmaceutical or industrial compounds into the food supply. The strategy should also identify criteria for determining when monitoring is no longer needed. In developing a strategy, the agencies should draw upon the analysis and conclusions of the National Research Council and the National Science and Technology Council.

USDA Response

USDA agrees, in part, with this recommendation. We support a risk-based approach to the oversight of **regulated** GE crops and any monitoring activities should also be risk-based in order to evaluate potential impacts on human health and the environment. However, our current regulations do not allow for APHIS to base its deregulation determinations upon any risks other than plant pest risks. Although USDA deregulation determinations are currently informed by the NEPA process which analyzes potential impacts on the human environment, ultimately USDA must deregulate a GE plant when it makes a determination that the GE plant is unlikely to be a plant pest. After USDA

See comment 1.

See comment 2.

See comment 3.

See comment 3.

See comment 4.

makes its deregulation determination, it would have no regulatory basis for further monitoring of the crop. If, however, USDA makes a partial deregulation determination, then we do agree that monitoring of the crop may be appropriate in specific circumstances where a potential for plant pest risk is identified. We likewise support having discussions with EPA and FDA regarding the possible parameters of USDA's involvement in plant pest risk-appropriate monitoring strategies for marketed deregulated GE crops should such monitoring be warranted in the future. USDA's overall approach towards monitoring was explained at length in a 2004 response to two National Research Council (NRC) reports, issued in 2000 and 2002. In that response USDA explained that an USDA deregulation is issued when the Agency has determined that the GE plant varieties do not pose a plant pest risk. This means that the Agency has found that unconfined release of these deregulated products is just as safe as that of their non-GE counterparts. Thus, once USDA has determined that the GE plant crop does not pose a plant pest risk, then it must remove the GE organism from its regulatory oversight. If USDA determined that monitoring of a product was required to mitigate a plant pest risk, USDA could choose to not deregulate the product or to partially deregulate it with stringent conditions including possible geographical limitations. Monitoring requirements could certainly be stipulated in permit conditions prior to the deregulation of any regulated GE crop.

USDA would like to also note that, as described in detail in the 2004 response to the NRC, there are many USDA programs involved in activities related to monitoring. In particular, USDA risk assessors help to establish priorities and review grants for the Biotechnology Risk Assessment Grant (BRAG) Program. BRAG is a competitive grant program administered by USDA's Cooperative State Research, Education, and Extension Service (CSREES) and the Agricultural Research Service (ARS). BRAG provides grants for research on the effects of introducing genetically modified organisms into the environment. CSREES and ARS also oversee other research programs related to risk assessment and risk mitigation of GE organisms. The response to the NRC also reported that in FY2003, ARS funded approximately $23 million in intramural environmental risk assessment research. Much of this intramural USDA research was conducted in the course of developing improved GE varieties. Approximately $4.4 million of this was specifically earmarked for risk assessment and risk mitigation research.

USDA also has several long-term programs that track many aspects of U.S. forest, crop, and range lands. The Forest Service's Forest Inventory and Analysis periodically gathers data on the status and trends in forest areas and locations across the United States.

Mr. Jim Jones
Page 4

plant diversity. In addition, the National Agricultural Statistics Service (NASS) conducts surveys and censuses of the Nation's 2.1 million farms. NASS focuses on the type of commodity produced, the yield, and the cost of production but also collects data on management practices. Since 2000, NASS has reported on state-by-state adoption rates of the major GE crops (soybeans, corn, and cotton) in major producing states and combined adoption rates for other states. Also, USDA's National Resource Conservation Service (NRCS) collects data on land use, and soil and water characteristics. Land use data are collected on 800,000 sites in the United States from satellite images every five years.

Sincerely,

Bruce I. Knight
Under Secretary
Marketing and Regulatory Programs

The following are GAO's comments on the U.S. Department of Agriculture's letter dated October 31, 2008.

GAO Comments

1. USDA commented that its regulatory system is science-based and limited to risks involving actual physical harm rather than perceived, but scientifically-unfounded, risks. In light of this comment, we have modified the wording of our draft recommendation to remove the reference to "perceived health concerns" and instead emphasize that the agreement would cover GE crops that present or are likely to present public health concerns.
2. USDA commented that its regulations do not allow for the Animal and Plant Health Inspection Service (APHIS) to base its deregulation determinations upon any risks other than plant pest risks. As USDA notes in its October 9, 2008 proposed rule, the agency's regulations could be grounded in more than just its authority to regulate GE crops as plant pests. The Plant Protection Act gives the Secretary of

Agriculture the authority to regulate to prevent the introduction or dissemination of noxious weeds. Noxious weeds are defined as any plant or plant product that can injure or cause damage to, among other things, crops, livestock, interests of agriculture, public health, or the environment. In this context, USDA could, for example, monitor marketed GE crops for their economic effects on other segments of agriculture.

3. USDA stated that after it makes a decision to deregulate a GE crop it has no regulatory basis for further monitoring. However, USDA maintains the authority under the Plant Protection Act to regulate a GE crop that it has granted deregulated status to if it obtains new information indicating that the crop is a plant pest. A coordinated inter-agency monitoring program would be one way of obtaining such information.
4. USDA listed several programs that it noted are related to monitoring. We did not review the programs that USDA mentioned, but we believe that they could provide useful monitoring data related to GE crops. We suggest that USDA incorporate them into the recommended coordinated strategy.

APPENDIX III. COMMENTS FROM THE DEPARTMENT OF HEALTH AND HUMAN SERVICES (FDA)

Note: GAO comments supplementing those in the report text appear at the end of this appendix.

DEPARTMENT OF HEALTH & HUMAN SERVICES OFFICE OF THE SECRETARY

Assistant Secretary for Legislation

Washington, DC 20201

OCT 21 2008

Lisa Shames, Director
Natural Resources and Environment
U.S. Government Accountability Office
441 G Street NW
Washington, DC 20548

Dear Ms. Shames:

Enclosed are the Department's comments on the U.S. Government Accountability Office's (GAO) draft report entitled: "Genetically Engineered Crops: Agencies Are Proposing Changes to Improve Oversight, but Could Take Additional Steps to Enhance Coordination and Monitoring" (GAO-09-60).

The Department appreciates the opportunity to review and comment on this report before its publication.

Sincerely,

Jennifer R. Luong for
Vincent J. Ventimiglia, Jr.
Assistant Secretary for Legislation

Attachment

See comment 1.

Food and Drug Administration General Comments on the Government Accountability Office's (GAO) Draft Report Entitled, "Genetically Engineered Crops-Agencies Are Proposing Changes to Improve Oversight, but Could Take Additional Steps to Enhance Coordination and Monitoring" (GAO 09-60)

The Food and Drug Administration (FDA) welcomes the Government Accountability Office's (GAO) draft report on bioengineered foods and appreciates the opportunity to review and provide comments. In addition to the FDA's responses to the recommendations, we have provided GAO with some technical comments regarding the draft report. FDA believes that its current processes for evaluating bioengineered foods and new proteins in such foods provides appropriate oversight and protection of the food and feed supplies. FDA also believes that it closely and effectively coordinates with the United States Department of Agriculture (USDA) and the Environmental Protection Agency (EPA) in the regulation of genetically engineered (GE) food plants. FDA does agree, however, that certain additional actions would increase transparency of, and enhance public confidence in, the agency's evaluation processes.

GAO RECOMMENDATIONS

To improve transparency and mitigate the impact of an unauthorized release into the food or feed supply of a regulated GE plant that has completed an early food safety evaluation, we recommend that the Commissioner of FDA fulfill the agency's commitment to post the results of completed early food safety evaluations on its Web site and add the results of future evaluations within 120 days of receiving the submission from the plant developer.

FDA RESPONSE

FDA agrees that posting the results of completed early food safety evaluations on our Web site would improve transparency. As described in the *Guidance to Industry: Recommendations for the Early Food Safety Evaluation of New Non-Pesticidal Proteins Produced by New Plant Varieties Intended for Food Use*, FDA has stated that the agency plans to respond to a complete submission within 120 days of receipt, and to make the text of that response letter easily accessible to the public via the Internet. FDA intends to make every effort to fulfill the commitments made in the guidance document; however, to date other activities of greater public health priority have been the focus of the agency's, and particularly, the Center for Food Safety and Applied Nutrition's, limited

See comment 2.
See comment 3.

<u>Food and Drug Administration General Comments on the Government Accountability Office's (GAO) Draft Report Entitled, "Genetically Engineered Crops-Agencies Are Proposing Changes to Improve Oversight, but Could Take Additional Steps to Enhance Coordination and Monitoring" (GAO 09-60)</u>

GAO RECOMMENDATION

To reduce the risk and impact of unauthorized releases, we recommend that the Secretary of Agriculture and the Commissioner of FDA develop a formal agreement to share information concerning GE crops with novel genetic traits that, if unintentionally released into the food or feed supply, could cause real or perceived health concerns and have negative financial consequences for the food and agriculture industry. With information from USDA about permits or notifications for field trials of such GE crops, FDA could identify which GE crops might benefit from an early food safety assessment and encourage the developers of those crops to participate in assessments. With assistance from FDA, USDA could make meaningful and transparent use of the health assessment data available through FDA's early food safety assessments in its risk assessment of GE crops.

FDA RESPONSE

FDA agrees in part with this recommendation. FDA agrees that it would be useful to explore possible mechanisms to facilitate information sharing with USDA to reduce the risk and potential public health impact of an unauthorized release of a GE crop. In the recent past, USDA and FDA have worked effectively to share information when an unauthorized release has occurred. However, the agency intends to explore development of a formal mechanism to facilitate these exchanges, recognizing that such a mechanism may make interagency coordination more transparent and thereby enhance public confidence.

The recommendation suggests that FDA obtain information about GE crops that, if unintentionally released into the food or feed supply, could cause "perceived health concerns." FDA uses a risk-based approach that focuses its resources on issues that either do or are likely to present public health concerns, as opposed to those issues that present only "perceived health concerns." FDA believes that focusing its resources on issues most likely to pose public health concerns provides the greatest public health protection. In addition, focusing resources on "perceived health concerns" would divert resources from activities providing significant public health protection to activities with little or no discernible public health benefit.

GAO also suggests that FDA obtain information about GE crops that, if unintentionally released into the food or feed supply, could have "negative financial consequences for the food and agriculture industry." While recognizing that the food and agriculture industry may experience negative financial consequences in the event of an unauthorized release of a GE crop, this issue falls outside the scope of FDA's mandate to protect and promote the public health.

See comment 4.

See comment 5.

Food and Drug Administration General Comments on the Government Accountability Office's (GAO) Draft Report Entitled, "Genetically Engineered Crops-Agencies Are Proposing Changes to Improve Oversight, but Could Take Additional Steps to Enhance Coordination and Monitoring" (GAO 09-60)

GAO RECOMMENDATION

To help ensure that unintended consequences arising from the marketing of GE crops are detected and minimized, we recommend that the Secretary of Agriculture, the Administrator of EPA, and the Commissioner of FDA develop a coordinated strategy for monitoring marketed GE crops and use the results to inform their oversight of these crops. Such a strategy should adopt a risk-based approach to identify the types of marketed GE crops that warrant monitoring, such as those with the greatest potential for affecting the environment or non-GE segments of agriculture, or those that might threaten food safety through the unintentional introduction of pharmaceutical or industrial compounds into the food supply. The strategy should also identify criteria for determining when monitoring is no longer needed. In developing a strategy, the agencies should draw upon the analysis and conclusions of the National Research Council and the National Science and Technology Council.

FDA RESPONSE

FDA agrees in part with this recommendation. FDA agrees that any monitoring strategy should adopt a risk-based approach. FDA does not believe that post-market monitoring of foods derived from GE crops currently on the market is necessary for the same reasons presented in the report (p. 35); there is no scientific evidence or even a hypothesis suggesting long-term harm from consumption of these foods. FDA would consider specific risk-based monitoring efforts should marketed GE crops intended for food or feed warrant such monitoring in the future. FDA intends to discuss coordination issues with USDA and EPA to be better prepared in case a situation should arise in the future that warranted such monitoring.

FDA notes GAO's concern regarding the potential for GE crops producing pharmaceutical or industrial substances to be inadvertently present in the food or feed supply (p. 35). As part of the coordinated effort to regulate such crops, USDA establishes strict permit conditions and performs rigorous inspections, including an increased rate of inspections, so that crops producing substances not intended for use in the food or feed supply do not inadvertently become part of the food or feed supply. FDA and USDA closely and effectively communicate and coordinate when there is a concern about the safety of the food or feed supply. FDA believes that random food product or commodity sampling to detect GE crops producing pharmaceutical or industrial substances in food and feed would present significant technical challenges and greatly affect resources, and would not likely be nearly as effective as USDA's current

The following are GAO's comments on the Department of Health and Human Service's letter dated October 21, 2008.

GAO Comments

1. FDA commented that it intends to make every effort to fulfill its commitment to post the results of its early food safety evaluations but that its focus has been on higher public health priorities. We recognize that FDA has competing priorities and finite resources, but we continue to believe that implementing this recommendation would be a relatively low-cost way to increase public transparency and trust and mitigate the impact of the unintended release of GE crops subject to early food safety evaluations.
2. FDA commented that it uses a risk-based approach that focuses its resources on issues that present or are likely to present public health concerns as opposed to issues that present only perceived health concerns. In light of this comment, as well as USDA's similar comment, we modified the wording of the recommendation to emphasize that the agencies develop a formal agreement to share information on GE crops that present or are likely to present public health concerns.
3. We acknowledge FDA's statement that the financial consequences of unintended releases fall outside it authority to protect and promote the public health. However, USDA, which bears some responsibility for promoting and expanding agricultural markets, may be concerned with these consequences. Thus, while we modified this recommendation to emphasize sharing information on GE crops that present or are likely to present health concerns, we also retained reference to the financial consequences of unintended releases. As we reported, known unintended releases of GE crops to the food and feed supply apparently have not caused health effects, but several led to financial losses.
4. FDA commented that it does not believe that post-market monitoring of foods derived from GE crops currently on the market is necessary. However, in making the recommendation that the agencies develop a coordinated monitoring strategy, our concern, in part, is the potential for GE crops producing pharmaceutical or industrial substances to be inadvertently present in the food or feed supply. Their presence could violate the Federal Food, Drug, and Cosmetic Act, could cause harm to human or animal health, and would likely cause financial harm to the agriculture and food industry. In light of this possibility, as well as the likelihood that the use of GE crops to produce these substances

will increase in the future, we believe that the agencies should develop a risk-based coordinated strategy to monitor for their presence in the food and feed supply.

5. FDA commented that USDA establishes strict permit conditions and performs inspections to minimize the likelihood that crops producing substances not intended for the food or feed supply might inadvertently become part of that supply, and that random FDA sampling to detect such substances would be difficult, expensive, and not as effective as USDA's actions. We acknowledge that USDA imposes strict permit conditions and requires frequent inspections of GE crops that produce pharmaceutical and industrial substances. However, while USDA may be able to reduce the likelihood of unintended releases of these crops, FDA has primary authority over the safety of the food and feed supply. Because biological substances such as GE crops are inherently difficult to control and there may be an increasing use of GE crops to produce an even wider array of pharmaceutical and industrial compounds in the future, we continue to believe that FDA and the other agencies should develop a risk-based coordinated strategy to monitor for their unintentional release. Furthermore, in the United States (1) GE crop varieties are grown extensively, (2) most processed foods contain ingredients from GE crops, and (3) genetic modifications are becoming increasingly complex in response to pressures to increase yields for food and biofuel. We believe these factors also argue for risk-based monitoring.

Appendix IV. U.S. Legal Framework for Regulation of GE Crops

On June 26, 1986, OSTP published the *Coordinated Framework* in the *Federal Register*.[22] The framework describes the comprehensive federal regulatory policy for ensuring the safety of biotechnology research and products. According to OSTP, existing statutes provide a basic network of agency jurisdiction over research and products and help ensure reasonable safeguards for the public. While OSTP recognized that the *Coordinated Framework* might need to evolve through administrative or legislative actions, it determined that existing laws would adequately address the regulatory needs for biotechnology.

The *Coordinated Framework* outlined the roles and responsibilities of relevant federal agencies, including USDA, EPA, FDA, the Occupational Safety and Health Administration, the National Institutes of Health, and the National Science Foundation. The framework also identified the relevant laws that would govern those agencies' activities regarding biotechnology. Table 4 contains summaries of key provisions in the primary laws that the agencies have used to regulate GE crops as well as a brief explanation of their relevance to biotechnology. Three of these laws—administered by USDA, EPA, FDA, or a combination of these agencies—include the Plant Protection Act; the Federal Insecticide, Fungicide, and Rodenticide Act; and the Federal Food, Drug, and Cosmetic Act. In addition, the table contains a summary of the relevant provisions of the National Environmental Policy Act of 1969; procedures outlined in that law must be followed by USDA, EPA, and FDA, when applicable.

Table 4. Key Legislation That Is Relevant to the Regulation of GE Crops

General application to GE organisms	Major relevant provisions
Legislation: Plant Protection Act[a]	
With respect to genetic engineering, USDA currently defines as a "regulated article" any organism that has been altered or produced through genetic engineering if the donor organism, recipient organism, or vector or vector agent belongs to any genera or taxa designated in a list of plant pests and meets the definition of plant pest or if the APHIS Administrator determines it is a plant pest or has reason to believe it is a plant pest. Statutory and regulatory requirements that apply to plant pests also apply to GE plants that meet the definition of plant pests. As described in this report, USDA is considering changes to its regulations that would also recognize the agency's authority to regulate GE plants as noxious weeds.	Repealed the Federal Plant Pest Act, the Plant Quarantine Act, and the Federal Noxious Weed Act of 1974, and several other related provisions. Defines a plant pest as any living stage of a protozoan, nonhuman animal, parasitic plant, bacterium, fungus, virus or viroid, infectious agent or other pathogen, or any article similar to or allied with the foregoing items. Prohibits the importation, entry, exportation, or movement of any plant pest in interstate commerce, unless authorized by the Secretary of Agriculture under permit. Authorizes the Secretary of the U.S. Department of Agriculture to allow importation, entry, exportation, or movement of a specific plant pest without a permit when he or she finds a permit is not necessary. Allows any person to petition to add or remove a plant pest from the list of plant pests exempt from the prohibition and directs the Secretary to act on the petition. Authorizes the Secretary to issue regulations to prohibit or restrict the importation, entry, exportation, or movement in interstate commerce of any plant product, biological control organism, noxious weed, article, or means of conveyance if he or she determines it is necessary to prevent the introduction or dissemination of a plant pest or noxious weed. Such regulations could include requiring permits or certificates of inspection. Authorizes the Secretary to publish, by regulation, a list of noxious weeds prohibited or restricted from entering the United States or that are subject to restrictions in interstate movement.

Table 4. (Continued)

General application to GE organisms	Major relevant provisions
	Authorizes the Secretary to, among other things, hold, seize, quarantine, or destroy any plant, plant pest, noxious weed, biological control organism,plant product, article, or means of conveyance, if he or she considers it necessary, to prevent the dissemination of a plant pest or noxious weed that is new or not widely prevalent in the United States, and is moving or has moved through the United States or interstate. States that no plant, plant pest, noxious weed, biological control organism, plant product, article, or means of conveyance shall be destroyed unless, in the opinion of the Secretary, there is no less drastic action that is feasible and that would be adequate to prevent the dissemination of any plant pest or noxious weed new or not widely prevalent in the United States. Authorizes the Secretary, upon a finding that an extraordinary emergency exists because of the presence of a plant pest or noxious weed that is new or not widely prevalent in the United States, to, among other things, hold, seize, quarantine, or destroy any plant, plant pest, noxious weed, biological control organism, plant product, article, or means of conveyance the Secretary has reason to believe is infested with the plant pest or noxious weed, or to quarantine any state or portion of a state in which the Secretary finds the plant pest or noxious weed. Authorizes the Secretary to inspect, without a warrant, any person or means of conveyance moving (1) into the United States to determine if the person or means of conveyance is carrying an article subject to the act; (2) in interstate commerce upon probable cause that the person or means of conveyance is carrying an article subject to the act; or (3) in intrastate

Table 4. (Continued)

General application to GE organisms	Major relevant provisions
	commerce within a state, portion of a state, or premises that is quarantined as part of an extraordinary emergency upon probable cause. Authorizes the Secretary to enter any premises in the United States for conductinginspections or seizures with a warrant issued upon probable cause that there is an article subject to the act on the premises. Grants the Secretary power to subpoena the attendance and testimony of witnesses, the production of all evidence, and the direction to permit inspections of premises related to the administration or enforcement of the act. Established criminal and civil penalties for violations of the act. Authorizes the Secretary to issue regulations and orders he or she considers necessary to carry out the act. Preempts states from regulating the movement in interstate commerce of any article subject to the act to control or eradicate a plant pest or noxious weed or to prevent the dissemination of a biological control organism, plant pest, or noxious weed if the Secretary has already issued a regulation, or to prevent the dissemination of the biological control organism, plant pest, or noxious weed within the United States.
Legislation: Federal Insecticide, Fungicide, and Rodenticide Act[b]	
With respect to genetically engineered organisms, EPA regulates the pesticides produced in plants, as well as the genetic material that produces such pesticides. These pesticides are known as "plant- incorporated protectants." The statutory and regulatory requirements that	Unless otherwise authorized by the act, prohibits the selling in any state of any pesticide that has not been registered under the act, and authorizes the Administrator of the Environmental Protection Agency to limit, by regulation, the distribution, sale, or use in any state of any pesticide that is not registered under the act or is not subject to an experimental use permit or an emergency exemption. Establishes procedures to register a pesticide

Table 4. (Continued)

General application to GE organisms	Major relevant provisions
apply to pesticides in general—such as those concerning registration, labeling, experimental use permits, inspections, and enforcement—also apply to plant- incorporated protectants produced in GE crops.	with EPA. Directs the Administrator to publish guidelines specifying the kinds of information required to support the registration of a pesticide. Establishes time frames and procedures for the Administrator to review and approve of registration applications. Permits any person to apply to the Administrator for experimental use permits for pesticides. Directs the Administrator to review those applications and either approve the permit or notify the applicant of the reasons for not issuing a permit. Limits experimental use permits to when the Administrator determines that the applicant needs such a permit to accumulate information necessary for registration of a pesticide under the act. Allows the Administrator to set a temporary tolerance level for pesticide residues before issuing an experimental use permit. Allows the Administrator, by regulation, to authorize states to issue experimental use permits for pesticides. Authorizes the Administrator to cancel a pesticide registration if it appears to the Administrator that the pesticide or its labeling does not comply with the act, or suspend registration to prevent an imminent hazard during the time proceedings are pending. Prohibits the production of pesticides subject to the act (or an active ingr-edient used in producing a pesticide subject to the act) unless the establishment at which such pesticides are produced is registered with the Administrator. Authorizes the Administrator to prescribe regulations requiring producers to maintain records with respect to their operations and the pesticides produced as the Administrator determines necessary for the effective enforcement of the act, and to make those records available

Table 4. (Continued)

General application to GE organisms	Major relevant provisions
	for inspection and copying. Requires producers to permit EPA, upon a valid request, access to and to copy all records showing delivery, movement, or holding of pesticides. Authorizes EPA to enter, at reasonable times, any establishment where pesticides are held for distribution or sale to inspect and obtain samples. EPA may obtain a warrant from a court of competent jurisdiction to enter and inspect an establishment or inspect and copy records when there is reason to believe that provisions of the act have been violated. States that it is unlawful for any person in any state to distribute or sell, among other things, any pesticide that is not registered (unless otherwise authorized under the act) or any pesticide that is misbranded or adulteryated. In addition, the act makes it unlawful for any person to, among other things, fail to prepare and maintain records, submit reports, or allow inspection under the act. Authorizes the Administrator to issue "stop sale, use, or removal" orders whenever a pesticide is found by the Administrator in any state and there is reason to believe—on the basis of inspections or tests—that the pesticide is in violation of the act, or intended to be sold or distributed in violation of the act. States that unsold pesticides (or pesticides in unbroken packages) being or having been transported, or sold or offered for sale in any state in violation of the act, shall be liable for seizure in any district court in a district where found if, among other things, the pesticide is misbranded or adulterated or is not registered under the act. The act also establishes civil and criminal penalties associated with violations of the act.

Table 4. (Continued)

General application to GE organisms	Major relevant provisions
	Authorizes the Administrator to exempt federal and state agencies from the provisions of the act if the Administrator determines emergency cond-itions exist that require such an exemption. Authorizes the Administ-rator to, by regulation or order, issue requirements and procedures for personswho store or transport pesticides the registration of which has been cance-led or suspended, for persons who dispose of stocks of pesticides the reg-istration of which has been suspended, and for the disposal of any pestic-ide the registration of which has been canceled. Directs the Administrator to order a recall of a pesticide, the registration of which has been suspend-ded or canceled, if he or she determines the recall is necessary to protect health or the environment. If the Administrator finds that voluntary recall by the registrant would be as safe and effective as a mandatory recall, the Administrator shall request a plan for that recall, and if the plan is adequ-ate, order the registrant to conduct the recall under the plan. Authorizes the Administrator to enter into cooperative agreements with states and Indian tribes to delegate the authority to cooperate in the enforcement of the act. Allows states to regulate the sale or use of federally registered pesticides in the state to the extent the regulation does not permit any sale or use prohibited by the act. Allows states to provide registration for additional uses of federally registered pesticides formulated for distribution and use within that state to meet special local needs in accordance with the purposes of the act. Authorizes the Administrator to prescribe regulations to carry out the provisions of the act, and establishes procedures for developing and

Table 4. (Continued)

General application to GE organisms	Major relevant provisions
	finalizing those regulations. Authorizes the Administrator to exempt, by regulation, any pesticide from the requirements of the act if he or she determines the pesticide is adequately regulated by another federal agency or, because of its character, it is unnecessary to be subject to the act in order to carry out the purposes of the act. Authorizes the Administrator, after notice and comment rulemaking, to, among other things, declare a pest, any plant or animal life (other than man or a bacteria, virus, or other micro-organism on or in living man or animals) that is injurious to health or the environment and establish standards with respect to the package, container, or wrapping in which a pesticide is enclosed. Declares that states shall have primary enforcement authority for pesticide use violations during a period for which the Administrator determines that the state has adopted adequate pesticide use laws and regulations, adopted and implemented adequate procedures for enforcement, and will keep records and reports showing the adoption of adequate laws and regulations and the adoption and implementation of adequate procedures. In addition, declares that states that have entered into cooperative agreements with the Administrator to receive delegated cooperative enforcement authority will have primary authority for enforcement, and that the Administrator will have primary enforcement authority for those states that do not. States that whenever the Administrator determines that a state with primary enforce-ement authority is not carrying out such responsibility, the Administrator will notify the state. After 90 days, if the Administrator determines the state's program remains inadequate, the Administrator can rescind, in whole or in part, the state's primary enforcement authority.

Table 4. (Continued)

General application to GE organisms	Major relevant provisions
Legislation: Federal Food, Drug, and Cosmetic Act, as amended[c]	
All domestic and imported foods and feeds under FDA's jurisdiction, whether or not they are derived from GE crops, must meet the same standards. Any food additive, including any introduced through genetic engineering, cannot be marketed before it receives FDA approval. However, substances added to foods that are "generally recognized as safe" under the conditions of intended use do not require FDA approval to be lawfully marketed. In 1992, FDA determined that most substances likely to become components of food as a result of genetic engineering would be the same or similar to substances commonly found in food. FDA encourages developers of GE foods to voluntarily notify the agency before marketing the foods. Notification leads to a consultation process between the agency and the company regarding the safety of the food in question.	Describes the mission of the Food and Drug Administration to, among other things, protect the public health by ensuring that foods are safe, wholesome, sanitary, and properly labeled. Defines "food" as articles used for food or drink for man or other animals, chewing gum, and articles used for components of any such articles. Defines "food additive" as any substance the intended use of which results or may reasonably be expected to result, directly or indirectly, in its becoming a component or otherwise affecting the characteristic of any food, if such substance is not "generally recognized as safe," as described in the act and implementing regulations, under the conditions of its intended use. However, pesticide chemicals and pesticide chemical residues, among other things, are not considered food additives. Prohibits the adulteration or misbranding of any food in interstate commerce, and the delivery for introduction into or receipt in interstate commerce of any adulterated or misbranded food. Gives U.S. district courts jurisdiction to, among other things, enjoin violations of the prohibitions, or to seize adulterated or misbranded food in interstate commerce. Provides criminal penalties for violations of these prohibitions. Authorizes FDA to temporarily detain food when there is credible evidence or information indicating that such article presents a threat of serious adverse health consequences or death. Deems a food to be "adulterated" when, among other things, the food bears or contains any poisonous or deleterious substance that may render it injurious to health; any pesticide chemical residue, unless the quantity of

Table 4. (Continued)

General application to GE organisms	Major relevant provisions
	the residue is within the limits of the tolerance set by EPA, or an exemption from the requirement of a tolerance is in effect; or any food additive that is unsafe. (Generally, food additives are considered unsafe unless a regulation is in effect prescribing the conditions under which it may be used safely.) Authorizes the Administrator of EPA to establish tolerances for pesticide chemical residues in or on food if he or she determines the tolerance is safe. Authorizes the Administrator to establish exemptions to required tolerances if he or she determines the exemption is safe. Requires those who manufacturer, process, pack, distribute, receive, hold, or import food (except farms and restaurants) to allow the Secretary of Health and Human Services (the Secretary) (delegated to FDA), when there is a reasonable belief that an article of food is adulterated and presents a threat of serious adverse health consequences or death, upon written notice at reasonable times and within reasonable limits and in a reasonable manner, to have access to and copy all records relating to such article of food needed to assist the Secretary (delegated to FDA) in determining whether the food is adulterated and presents such a threat. Authorizes the Secretary (delegated to FDA) to establish, by regulation, requirements regarding the establishment and maintenance of records needed to identify the immediate previous sources and immediate subsequent recipients of food to address credible threats of serious health consequences or death. Directs the Secretary (delegated to FDA) to, by regulation, require facilities that manufacture, process, pack, or hold food for consumption (not including farms, restaurants, other retail food

Table 4. (Continued)

General application to GE organisms	Major relevant provisions
	establishments, certain nonprofit food establishments, or certain fishing vessels) to register with the Secretary. Authorizes the Secretary (delegated to FDA) to promulgate regulations for the efficient enforcement of the act, and to conduct examinations and investigations for the purposes of the act. Authorizes the Secretary (delegated to FDA), for the purposes of enforcement and upon written notice, to enter any factory, warehouse or establishment in which food is manufactured, processed, packed, or held for introduction in interstate commerce or after such introduction, or to enter any vehicle being used to transport or hold such food in interstate commerce. Authorizes the Secretary (delegated to FDA), for the purposes of enforcement and upon written notice, to inspect, at reasonable times, and within reasonable limits, and in a reasonable manner, any factory, warehouse, establishment, or vehicle and all pertinent equipment, finished and unfinished materials, containers, and labeling therein. Authorizes the refusal of admission of any article of food if, among other reasons, it appears from examination of samples or otherwise that such article is adulterated or misbranded. Directs FDA, under certain circumstances and upon credible evidence or information indicating that an article of food presents a threat of serious adverse health consequences or death, to request the Secretary of Treasury to hold such article for up to 24 hours to enable the Secretary (delegated to FDA) to inspect, examine, or investigate such article.

Table 4. (Continued)

General application to GE organisms	Major relevant provisions
Legislation: National Environmental Policy Act of 1969[d]	
This act requires agencies with oversight responsibility for GE crops to consider the likely environmental effects of actions they are proposing, and if those actions would significantly affect the environment, provide an environmental impact statement. Such statements could be required for actions related to the regulation of GE crops. For example, USDA's effort to change its biotechnology regulations is being conducted under the provisions of the act. USDA also conducts environmental analyses when it receives a petition to grant nonregulated status to GE crops.	Directs all federal agencies to include a detailed statement of, among other things, the environmental impact, adverse environmental effects that cannot be avoided, and alternatives to the proposed action in every recommenddation or report on proposals for major federal actions significantly affecting the quality of the human environment. Directs federal agencies to study, develop, and describe appropriate alternatives to recommended courses of action in any proposal that involves unresolved conflicts concerning alternative uses of available resources.

Source: GAO analysis of relevant provisions of these four statutes.

[a]Pub. L. No. 106-224, Tit. IV, §§ 401-442, 114 Stat. 438 (codified as amended at 7 U.S.C. §§ 7701- 7786).

[b]Pub. L. No. 80-104, 61 Stat. 163 (1947) (codified as amended at 7 U.S.C. §§ 136-136y).

[c]Pub. L. No. 75-717, 52 Stat. 1040 (1938) (codified as amended at 21 U.S.C. §§ 321-399).

[d]Pub. L. No. 91-190, 83 Stat. 852 (codified as amended at 42 U.S.C. §§ 4321 -4370f).

Appendix V. Summary Information on USDA Notifications and Permits for Field Trials of GE Crops

USDA acknowledges notifications submitted by developers seeking to import, move interstate, or conduct field trials of regulated GE material, including GE crops, or issues permits. For field trials involving low-risk GE materials, such as engineering a well-known protein into a new plant variety, the more streamlined notification process may be used, assuming other regulatory criteria are met. However, for higher-risk items, such as engineering an unfamiliar protein into a new plant, a permit may be required that provides, among other things, more specific conditions for containment of the GE crop during the field trial.

From 1987 through 2007, USDA approved over 13,000 notifications and permits for field trials. Over 90 percent of these approvals were notifications. Figure 3 shows the number of notifications and permits that USDA approved annually during this period.

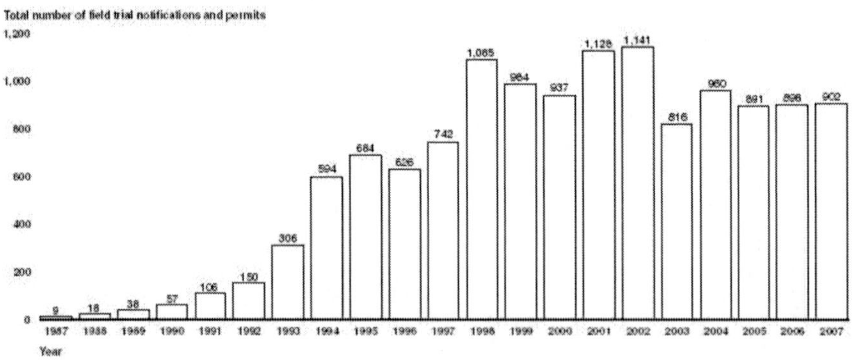

Source: USDA's Animal and Plant Health Inspection Service.

Note: USDA created the notification process in 1993; prior to that year, it only issued permits.

Figure 3. Annual Number of Notifications and Permits Approved by USDA, 1987 through 2007

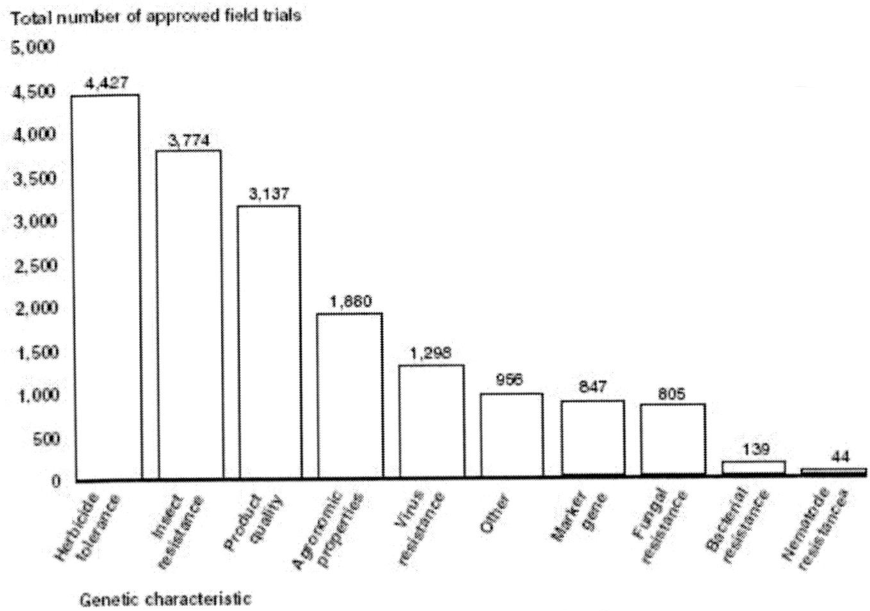

Source: Virginia Tech's Information Systems for Biotechnology.
[a]Nematodes are wormlike organisms.

Figure 4. Number of Field Trials by the Genetic Characteristic Tested, 1987 through 2007

Over time, GE crop developers have conducted field trial experiments on a variety of characteristics engineered into plants. The characteristics tested most often have been herbicide tolerance and insect resistance. Figure 4 presents data on the number of field trials by the type of characteristic being tested from 1987 through 2007. Because developers may test more than one characteristic during a field trial, the total number of characteristics (over 17,000) exceeds the more than 13,000 approved notifications and permits.

APPENDIX VI. INFORMATION ON THE COMMERCIALIZATION OF GE CROPS

If field tests of a new GE crop are successful—for example, the desired trait, such as herbicide resistance, is expressed, and there are no unresolved safety issues—developers may seek to commercialize the product. In general, to do this, developers must petition USDA to deregulate the GE crop. In turn, to grant this "nonregulated" status, USDA must determine that the crop does not constitute a "plant pest." For GE crops engineered to include a pesticidal protectant, the developer must also obtain a pesticide registration from EPA. Finally, prior to introducing GE crops into the food or feed supply, developers are encouraged to consult with FDA on the crops' potential allergenicity, toxicity, and antinutrient (interference with nutrient absorption) properties. Assuming these regulatory agencies do not act to restrict the growth or use of a GE crop, it can enter into the food or feed supply and mix with conventional (non-GE) varieties without being monitored, traced, or labeled.

USDA's Deregulation of GE Crops

Within USDA, APHIS bears the main responsibility for assessing the environmental safety of GE crops. The primary focus of this agency's review is to determine whether a plant produced through biotechnology is a plant pest. Developers can petition the agency to exempt a GE plant from regulation once they have collected sufficient and appropriate data regarding the potential environmental impact of a GE plant. The agency may choose to grant the petition in whole or in part or to deny the petition. As of July 14, 2008, USDA had received 113 petitions to deregulate regulated GE crops, of which it approved 73. (See table 5 at the end of this appendix for information regarding the GE crops addressed by those 73 petitions.) Of the remainder, 12 were pending, 27 were withdrawn by the petitioner, and USDA identified 1 as "incomplete." In general, USDA approval of a petition to deregulate allows the developer to market the product in the United States.

EPA's Registration of GE Pesticides Known as Plant-Incorporated Protectants

EPA is responsible for regulating genetic modifications in plants that protect them from insects, bacteria, and viruses, as well as the genetic material that causes the pesticide to be produced. These protectants are subject to the agency's regulations on the sale, distribution, and use of pesticides. In order for field-testing of plants with such protectants to be performed on more than 10 acres of land, EPA must grant an experimental use permit. Prior to commercialization of a GE plant with such a protectant, EPA reviews the application for approval of the protectant, solicits public comments, and may seek the counsel of external scientific experts. For registrations of new GE pesticides, EPA routinely examines information regarding the identification of the new genetic material, toxicity or allergenicity concerns, and possible effects on wildlife. EPA also evaluates whether the residues of the pesticide in food will be "safe" and determines whether a tolerance or tolerance exemption can be issued. Since 1995, EPA has registered 29 GE pesticides engineered into 3 crops— corn, cotton, and potatoes—5 of which have since been voluntarily canceled.[23] All currently registered GE pesticides have received an exemption from the requirement of a tolerance, indicating EPA's determination that any level of pesticidal residue from these crops is safe for food and feed.

FDA's Voluntary Consultation Process for GE Food and Feed Crops

FDA has primary responsibility for ensuring the safety of most of the nation's food supply. The Federal Food, Drug, and Cosmetic Act prohibits the adulteration of food in interstate commerce.[24] In this context, FDA encourages companies developing GE foods to voluntarily notify the agency before marketing the foods. Notification leads to a two-part consultation process between the agency and the company that initially involves discussions of relevant safety issues, and subsequently involves the company's submission of a safety assessment report containing a summary of test data on the food in question.[25] The purpose of these test data is to demonstrate that the GE food item presents no greater risk of allergenicity, toxicity, or antinutrient properties than its conventional counterparts. At the end of the consultation, FDA evaluates the data and may send a letter to the company stating that the agency has no further questions, indicating in effect that it sees no reason to prevent the company from commercializing the GE food. Although this consultation is voluntary, FDA officials said that they are not aware of any GE food or feed products intentionally marketed to date that have not gone through the consultation process. As of July 2008, FDA had completed 72 voluntary consultations on GE crops intended for use in human food, animal feed, or both; not all of these items were marketed.

FDA also has regulatory authority over pharmaceutical products derived from GE crops. These products may be marketed—regardless of whether the associated GE crops have been deregulated by USDA—only with FDA approval of a marketing application. However, as of July 2008, FDA had not received any applications to market pharmaceutical products from GE crops.[26]

Many GE Crops Have Been Marketed for a Variety of Purposes

Many GE crops have been marketed in the United States and other countries for a variety of purposes, such as food or feed use. For example, in the United States, GE varieties accounted for about 80 percent of the corn, 92 percent of the soybeans, and 86 percent of the cotton planted in 2008. Furthermore, according to food industry sources, over 70 percent of the processed foods sold in the United States contain ingredients and oils from GE crops.

However, not all GE crops have been marketed in the United States, and others were marketed for several years but then were withdrawn from commercial production. Some of the GE crops marketed in the United States may also be approved for marketing in other countries. In some instances, those countries have placed restrictions on the use of these crops. Table 5 provides information on GE crops granted nonregulated status by USDA, their approved uses in the United States and other countries, and their marketing status in the United States.

Table 5. GE Crops Granted Nonregulated Status by USDA and their Marketing Status in the United States and Other Countries

GE crop/ Institution	Genetic transformation, or "event,"[a] and trait[b]	Commercialized in the United States[c]	Commercial name[d]	Approved for:				
				All uses	Environment	Planting[e]	Food	Feed
Canola								
Aventis	MS1&RF1 MS1&RF2 (PGS1) Trait: HT+F	Last sold in 2003	SeedLink Canola	United States, Australia, Canada, European Union	Japan		China, Japan, New Zealand, South Africa, South Korea	China, Japan, South Africa
Aventis	Topas 19/2 (HCN92, HCN10) Trait: HT	Last sold in 2003	[d]	United States, Canada	Australia, Japan	Australia	Australia, China, European Union, Japan, Mexico, New Zealand (HCN92), South Africa, South Korea	China, European Union, Japan, South Africa
AgrEvo	MS8xRF3 Trait: HT+F	Yes	[d]	United States, Australia, Canada, Japan	European Union, South Korea		China, European Union, Mexico, New Zealand,	China, European Union, Mexico, South Africa

Table 5. (Continued)

GE crop/ Institution	Genetic transformation, or "event,"[a] and trait[b]	Commercialized in the United States[c]	Commercial name[d]	Approved for:				
				All uses	Environment	Planting[e]	Food	Feed
							South Africa, South Korea	
AgrEvo	T45 (HCN28) Trait: HT	Last sold in 2005	LibertyLink® Canola	United States, Australia, Canada	Japan, South Korea		China, European Union, Japan, Mexico, New Zealand, South Korea	China, European Union, Japan
Calgene	PCGN 3828-212/86–18 PCGN 3828-212/86–23 (23-18-17, 23-198) Trait: OC	PCGN 3828-212-86-23 (last sold in 1998) No: PCGN 3828-212-86-18	Laurical	United States, Canada				
Monsanto	GT200 (RT200) Trait: HT	No	Westar Roundup Ready®	United States	Canada, Japan		Canada, Japan	Japan
Monsanto	GT73 RT73 Trait: HT	Yes	Roundup Ready Canola® or Westar Roundup Ready®	United States, Canada, Japan	Australia, European Union, South Korea	Australia	Australia, China, European Union, Mexico, New Zealand,	China, European Union, Mexico, Philippines

Table 5. (Continued)

GE crop/ Institution	Genetic transformation, or "event,"[a] and trait[b]	Commercialized in the United States[c]	Commercial name[d]	Approved for:				
				All uses	Environment	Planting[e]	Food	Feed
							Philippines, South Korea	
Chicory								
Bejo	RM3-3 RM3-4 RM3-6 Trait: HT+F	[c]		United States	European Union	European Union		
Cotton								
Aventis	LLCotton25 Trait: HT	Yes	LibertyLink Cotton	United States	South Korea		Australia, Canada, Japan, Mexico, New Zealand, South Korea	Canada, Japan, Mexico
Calgene	31807/31808 Trait: HT+IR	[c]		United States	Japan		Canada, Japan	Japan
Calgene	BXN Trait: HT	[c]		United States	Japan		Australia, Canada, Japan, Mexico, New Zealand	Australia, Canada, Japan
Mycogen/ Dow	281-24-236 Trait: IR	No		United States			Canada, Mexico	Canada
Mycogen/ Dow	3006-210-23 Trait: IR	No		United States			Canada, Japan, Mexico	Canada, Mexico

Table 5. (Continued)

GE crop/ Institution	Genetic transformation, or "event,"[a] and trait[b]	Commercialized in the United States[c]	Commercial name[d]	Approved for:				
				All uses	Environment	Planting[e]	Food	Feed
DuPont	19-51A Trait: HT	No		United States				
Monsanto	MON1445 MON1698 Trait: HT	Yes: MON1445 No: MON1698	Roundup Ready®	United States, Argentina (MON1445), Columbia (MON1445)	Australia (MON1445), Japan, Mexico, South Africa	Australia, Mexico, South Africa	Australia, Canada, China, European Union (MON1445), Japan, Mexico, New Zealand, Philippines	Canada, China, European Union (MON1445), Japan, Philippines
Monsanto	MON15985 Trait: IR	Yes	Bollgard II®	United States, India, South Africa	Australia, South Korea	Australia	Australia, Canada, European Union, Japan, Mexico, New Zealand, Philippines, South Korea	Canada, European Union, Japan, Philippines
Monsanto	MON531 MON757 MON1076 Trait: IR	Yes: MON531 No: MON1076 MON757[c]	Bollgard®	United States, Argentina (MON531), Australia (MON531),	Indonesia, Japan, South Korea (MON531)	Indonesia	Canada, European Union (MON531), Japan, New Zealand,	Canada, European Union (MON531), Japan, Philippines

Table 5. (Continued)

GE crop/ Institution	Genetic transformation, or "event,"[a] and trait[b]	Commercialized in the United States[c]	Commercial name[d]	Approved for:				
				All uses	Environment	Planting[e]	Food	Feed
				Brazil, China, Colombia (MON531), India (MON531), Mexico, South Africa (MON531)			Philippines (MON531), South Korea (MON531)	(MON531)
Monsanto	MON88913 Trait: HT	Yes	Roundup Ready® Flex	United States, South Africa	Australia	Australia	Australia, Canada, Japan, Mexico, New Zealand, Philippines, Singapore, South Korea	Canada, Japan, Philippines
Syngenta Seeds	COT102 Trait: IR	[c]					United States, Australia, New Zealand	United States
Flax, Linseed								
University of Sask-atchewan	FP967 Trait: HT	[c]	CDC Triffid	United States, Canada				

Table 5. (Continued)

GE crop/ Institution	Genetic transformation, or "event,"[a] and trait[b]	Commercialized in the United States[c]	Commercial name[d]	Approved for:				
				All uses	Environment	Planting[e]	Food	Feed
Maize (corn)								
AgrEvo	CBH-351 Trait: HT+IR	Last sold in 2000	StarLink		United States	United States		United States
Plant Genetic Systems	MS3 Trait: HT+F	[c]		United States, Canada				
AgrEvo	MS6 Trait: HT+F	[c]		United States				
AgrEvo	T14 T25 Trait: HT	Yes: T25 T14: Last sold in 1999	Liberty Link™	United States, Argentina, Canada, European Union (T25), Japan (T25)	Japan (T14), South Korea (T25)		Australia (T25), China (T25), Japan (T14), Mexico, New Zealand (T25), Philippines (T25), Russia (T25), South Korea (T25), Taiwan (T25)	Australia (T25), China (T25), Japan (T14), Mexico, Philippines (T25), Taiwan (T25)
Dekalb Genetics Corpor-ation	B16 (DLL25) Trait: HT	Last sold in 1999	[d]	United States, Canada, Japan			Philippines, South Korea, Taiwan	Philippines, Taiwan

Table 5. (Continued)

GE crop/ Institution	Genetic transformation, or "event,"[a] and trait[b]	Commercialized in the United States[c]	Commercial name[d]	Approved for:				
				All uses	Environment	Planting[e]	Food	Feed
Dekalb Genetics Corpo-ration	DBT418 Trait: HT+IR	Last sold in 1999	Bt Xtra™	United States, Canada	Argentina, Japan		Australia, Japan, New Zealand, Philippines, South Korea, Taiwan	Japan, Philippines, Taiwan
Dow AgroScie-nces LLC	DAS-06275-8 (TC6275) Trait: IR	No		United States, Canada			Japan	
Dow Agro Sciences LLC/ Pioneer Hi-Bred Internation al Inc.	59122 (DAS-59122-7, Event 22) Trait: HT+IR	Yes	Herculex® RW	United States, Canada, Japan	European Union		Australia, European Union, Mexico, New Zealand, Philippines, South Korea, Taiwan	European Union, Mexico, Philippines, Taiwan
Monsanto	GA21 Trait: HT	Yes	Roundup Ready®	United States, Argentina, Canada, Japan	South Korea		Australia, China, European Union, Mexico, New Zealand, Philippines, Russia,	China, European Union, Philippines, Russia, South Africa, Taiwan

Table 5. (Continued)

GE crop/ Institution	Genetic transformation, or "event,"[a] and trait[b]	Commercialized in the United States[c]	Commercial name[d]	Approved for:				
				All uses	Environment	Planting[e]	Food	Feed
							South Africa, South Korea, Taiwan	
Monsanto	LY038 Trait: LYS	[c]		United States, Canada	Japan	Japan	Australia, Japan, Mexico, Philippines	Australia, Philippines
Monsanto	MON80100 Trait: IR	[c]		United States				
Monsanto	MON802 Trait: HT+IR	[c]	Yieldgard®	United States, Canada	Japan			
Monsanto	MON810 Trait: IR	Yes	Yieldgard®	United States, Argentina, Canada, European Union, Honduras, Japan, Philippines, South Africa, Uruguay	Colombia, South Korea		Australia, China, Colombia, Mexico, New Zealand, Russia, South Korea, Switzerland, Taiwan	China, Colombia, Russia, Switzerland, Taiwan
Monsanto	MON863 Trait: IR	Yes	[d]	United States, Canada,	South Korea		Australia, China,	China, European Union,

Table 5. (Continued)

GE crop/ Institution	Genetic transformation, or "event,"[a] and trait[b]	Commercialized in the United States[c]	Commercial name[d]	Approved for:				
				All uses	Environment	Planting[e]	Food	Feed
				Japan			European Union, Mexico, New Zealand, Philippines, Russia, Singapore, South Korea, Taiwan	Philippines, Russia, Singapore, Taiwan
Monsanto	MON88017 Trait: HT+IR	Yes	[d]	United States, Canada	Japan	Japan	Australia, Japan, Mexico, New Zealand, Philippines, South Korea, Taiwan	Mexico, Philippines, Taiwan
Monsanto	MON89034 Trait: HT+IR	[c]		United States	Canada, Japan		Japan	Canada, Japan
Monsanto	NK603 Trait: HT	Yes	Roundup Ready®	United States, Argentina, Canada, Japan, Philippines	South Korea, Uruguay	Uruguay	Australia, China, Colombia, European Union, Mexico, Russia,	China, Colombia, European Union, Russia, Singapore,

Table 5. (Continued)

GE crop/ Institution	Genetic transformation, or "event,"[a] and trait[b]	Commercialized in the United States[c]	Commercial name[d]	Approved for:				
				All uses	Environment	Planting[e]	Food	Feed
				,South Africa			New Zealand, Singapore, South Korea, Taiwan, Thailand	Taiwan, Thailand
Mycogen (Dow AgroSciences); Pioneer (Dupont)	1507 (TC1507) Trait: HT+IR	Yes	Herculex® I	United States, Argentina, Canada, Japan	Colombia, Uruguay	Uruguay	Australia, China, Colombia, European Union, Mexico, New Zealand, South Korea, Philippines, South Africa, Taiwan	China, Colombia, European Union, Philippines, South Africa, Taiwan
Pioneer Hi-Bred International Inc.	676 678 680 Trait: HT+F	No		United States				
Monsanto	MON809 Trait: HT+IR	No		United States, Canada	Japan			Japan

Table 5. (Continued)

GE crop/ Institution	Genetic transformation, or "event,"[a] and trait[b]	Commercialized in the United States[c]	Commercial name[d]	Approved for:				
				All uses	Environment	Planting[e]	Food	Feed
Northrup King	Bt11 Trait: HT+IR	Yes	[d]	United States,	Japan		Australia, China,	Australia, China,
				Argentina, Canada, Philippines, South Africa, Uruguay			European Union, Japan, Mexico, New Zealand, Russia, South Korea, Switzerland, Taiwan	European Union, Japan, Mexico, Switzerland, Taiwan
Syngenta Seeds	MIR604 Trait: IR	Yes	Agrisure RW Rootworm-Protected Corn		United States, Japan, Philippines	Japan	United States, Australia, Japan, Mexico, New Zealand, Philippines, South Korea	Mexico, Philippines
Ciba Seeds	176 (Bt 176) Trait: HT+IR	Yes	NaturGard™ KnockOut™	United States, Argentina, Australia, Canada, European Union	Japan		China, Japan, New Zealand, Philippines, South Africa, Switzerland,	China, Japan, Philippines, South Africa, Switzerland, Taiwan

Table 5. (Continued)

GE crop/ Institution	Genetic transformation, or "event,"[a] and trait[b]	Commercialized in the United States[c]	Commercial name[d]	Approved for:				
				All uses	Environment	Planting[e]	Food	Feed
							South Korea, Taiwan	
Papaya								
Cornell University	55-1/63-1	Yes	SunUp,	United			Canada	
	Trait: VR		Rainbow	States				
Plum								
USDA-Agricul-tural Research Service	ARS-PLMC5-6(C5) Trait: VR	[c]			United States			
Potato								
Monsanto	BT6 BT10 BT12 BT16 BT17 BT18 BT23 Trait: IR	BT6: Last sold in 2001 No: BT10, BT12, BT16, BT17, BT18, and BT23	Russet Burbank NewLeaf®	United States, Canada			Japan, Mexico, Philippines (BT16)	Mexico, Philippines (BT16)
Monsanto	ATBT04-6 ATBT04-27 ATBT04-30 ATBT04-31 ATBT04-36 SPBT02-5	ATBT04-6: Last sold in 2000 ATBT04-31: Last sold in 2000 ATBT04-36: Last sold in 2000 SPBT02-5: Last	Atlantic and Superior NewLeaf®	United States, Canada	Russia (SPBT02-5)		Australia, Japan, New Zealand, Philippines (SPBT02-5), Russia	Australia, Philippines (SPBT02-5)

Table 5. (Continued)

GE crop/ Institution	Genetic transformation, or "event,"[a] and trait[b]	Commercialized in the United States[c]	Commercial name[d]	Approved for:				
				All uses	Environment	Planting[e]	Food	Feed
	SPBT02-7 Trait: IR	sold in 2001 SPBT02-7[c] No: ATBT04-27, ATBT04-30					(SPBT02-5), South Korea (SPBT02-5)	
Monsanto	RBMT22-082 Trait: IR+VR	Last sold in 2000		United States, Canada			Australia, Japan, Mexico, New Zealand	Australia
Monsanto	RBMT21-129 RBMT21-350 Trait: IR+VR	RBMT21-350: Last sold in 2000 RBMT21-129: Last sold in 2000	Russet Burbank NewLeaf® Plus	United States, Canada			Australia, Japan, Mexico, New Zealand, Philippines, South Korea	Australia, Philippines
Monsanto	RBMT15-101 SEMT15-02 SEMT15-15 Trait: IR+VR	RBMT15-101: Last sold in 2001 SEMT15-02[c] SEMT15-15[c]	NewLeaf® Y	United States, Canada			Australia, Japan, Mexico, New Zealand, Philippines, South Korea	Australia, Mexico, Philippines
Rice								
AgrEvo	LLRICE06 LLRICE62 Trait: HT	No	Liberty Link®	United States			Canada, Mexico, Russia (LLRICE62)	Canada, Mexico

Table 5. (Continued)

GE crop/ Institution	Genetic transformation, or "event,"[a] and trait[b]	Commercialized in the United States[c]	Commercial name[d]	Approved for:				
				All uses	Environment	Planting[e]	Food	Feed
Bayer Crop-Science	LLRICE601 Trait: HT	No			United States	United States		
Soybean								
AgrEvo	A2704-12 A2704-21 A5547-35 W62 W98 Trait: HT	No	Liberty Link®	United States (all 5)	Canada (A2704-12), Japan (A2704-12)		Australia (A2704-12, A2704-21, A5547-35), Canada (A2704-12), European Union (A2704-12), Japan (A2704-12), Mexico (A2704-12, A2704-21, A5547-35), New Zealand (A2704-12, A2704-21, A5547-35), Russia (A2704-12), South Africa (A2704-12)	Canada (A2704-12), European Union (A2704-12), Japan (A2704-12), Mexico (A2704-12, A2704-21, A5547-35), South Africa (A2704-12)

Table 5. (Continued)

GE crop/ Institution	Genetic transformation, or "event,"[a] and trait[b]	Commercialized in the United States[c]	Commercial name[d]	Approved for:				
				All uses	Environment	Planting[e]	Food	Feed
AgrEvo	A5547-127 Trait: HT	No	Liberty Link®	United States	Japan		Japan, Mexico, Russia	Japan, Mexico
AgrEvo	GU262 Trait: HT	No		United States				
DuPont Canada Agri-cultural Products	G94-1 G94-19 G168 Trait: OC	c		United States, Canada	Japan		Australia, Japan, New Zealand	Japan
Monsanto	GTS40-3-2 Trait: HT	Yes	d	United States, Argentina, Brazil, Canada, Japan, Mexico, Paraguay, Romania, Uruguay	South Korea		Australia, China, Czech Republic, European Union, Malaysia, New Zealand, Philippines, Russia, South Korea, Switzerland, Taiwan, Thailand	China, Colombia, Czech Republic, European Union, Malaysia, Philippines, Russia, Switzerland, Taiwan, Thailand
Monsanto	MON89788 Trait: HT	No	Roundup Ready 2 Yield®	United States, Canada	Japan, Philippines, Taiwan		Japan, Philippines, Taiwan	Japan, Philippines, Taiwan

Table 5. (Continued)

GE crop/ Institution	Genetic transformation, or "event,"[a] and trait[b]	Commercialized in the United States[c]	Commercial name[d]	Approved for:				
				All uses	Environment	Planting[e]	Food	Feed
Squash								
Asgrow (United States); Seminis Vegetable Inc. (Canada)	CZW-3 Trait: VR	Yes	[d]	United States			Canada	
Upjohn (Seminis Vegetable Seeds)	ZW20 Trait: VR	Yes	[d]	United States			Canada	
Sugarbeet								
AgrEvo	T120-7 Trait: HT	No		United States, Canada			Japan	Japan
Monsanto	H7-1 Trait: HT	Yes	[d]	United States, Canada			Australia, European Union, Mexico, New Zealand, Philippines, Russia, Singapore, South Korea	European Union, Philippines, Singapore

Table 5. (Continued)

GE crop/ Institution	Genetic transformation, or "event,"[a] and trait[b]	Commercialized in the United States[c]	Commercial name[d]	Approved for:				
				All uses	Environment	Planting[e]	Food	Feed
Novartis Seeds; Monsanto	GTSB77 Trait: HT	No	InVigor™	United States			Australia, Japan, New Zealand,Phil ippines, Russia	Australia, Philippines
Tobacco								
Vector	Vector21-41 Trait: NIC	c			United States	United States		
Tomato								
Agritope Inc.	351N Trait: DR	c		United States				
Calgene	FLAVR SAVR[f] Trait: DR	c	FLAVR SAVR™	United States	Japan, Mexico		Canada, Japan, Mexico	Japan, Mexico
Calgene	N73 1436-111 Trait: DR	Last sold in 1997	FLAVR SAVR™	United States				
DNA Plant Techno-logy	1345-4 Trait: DR	c		United States			Canada, Mexico	
Corpora-tion Monsanto	5345 Trait: IR	No		United States			Canada	
Monsanto	8338 Trait: DR	No		United States				
Zeneca + Petoseed	B, Da, F Trait: DR	c		United States			Canada, Mexico	

Source: GAO analysis of data from USDA, EPA, FDA, the Biotechnology Industry Organization, the AGBIOS Company, and the International Service for the Acquisition of Agri-Biotech Applications.

[a] Some events have a synonymous name; those event names are shown in parentheses.

[b] HT (herbicide tolerance), IR (insect resistance), VR (virus resistance), DR (delayed ripening/altered shelf life), OC (modified oil content), LYS (enhanced lysine content), NIC (nicotine reduction), and F (fertility restored).

[c] In some cases, we were not able to determine from the cited sources whether the GE crop had been marketed.

[d] In some cases, we were not able to determine from the cited sources whether the GE crop had a specific commercial name other than its event name.

[e] Has been approved for planting/cultivation, but is not necessarily in commercial production at the present time.

[f] Thirty-three lines of FLAVR SAVR™ tomato were granted nonregulated status by USDA.

Appendix VII. Six Documented Incidents of Unauthorized Release of GE Crops into the Food and Feed Supply

As of August 2008, there were six documented incidents of the unauthorized release of GE crops into the food or feed supply, or into crops meant for the food or feed supply. Although federal agencies determined that these incidents did not harm human or animal health, they did cause financial losses in some cases, primarily from lost sales to countries that would not accept food or feed containing any amount of regulated GE varieties. The six incidents are discussed in the following text.

StarLink Corn – 2000

The first known unauthorized release of GE crops into the food supply occurred in 2000 and involved a GE corn variety known by its trademark name, StarLink. StarLink was engineered for insect resistance and herbicide tolerance by Aventis Crop Science. USDA deregulated StarLink in 1998, and FDA accepted Aventis' data showing that, other than its new pesticidal protein, StarLink was essentially the same as other commercially available corn varieties. However, EPA granted only a "split- registration" to the pesticidal protein in StarLink corn, thereby allowing residue of the protein in animal feed but not allowing it in the human food supply because of concerns that it may be an allergen.[27] In 2000, trace amounts of StarLink corn were found in commercially available taco shells. According to USDA and other sources, StarLink corn intended for animal feed, as well as corn grown in adjacent fields that cross-pollinated with StarLink, likely became commingled with corn approved for human consumption during harvesting, transportation, and storage.

Federal agencies took a number of actions to divert Starlink corn from the food supply. For example, APHIS began purchasing bushels of StarLink corn at a 25-cent premium, with Aventis agreeing to reimburse the agency for the costs. In addition, the food industry initiated recalls of over 300 products that could have contained the regulated protein. FDA also issued guidance for sampling and testing corn for the presence of this protein. These actions dramatically reduced the amount of the protein in the food supply. USDA testing done in 2006 and 2007 found no trace of the protein in the samples tested.

The StarLink incident had financial consequences, particularly in major U.S. export markets. In 2001, USDA reported that corn sales to Japan—the largest importer of U.S. corn—were down more than 20 percent from the previous year, and exports to South Korea were down more than 70 percent, although USDA noted that some of this drop resulted from other factors, such as larger-than-expected corn production and exports from Argentina and Brazil. One study estimated that the StarLink incident resulted in $26 million to $288 million in lost revenue for producers in market year 2000/2001.[28] (U.S. cash receipts for corn totaled about $15.2 billion in 2000.) In addition, this study estimated that the federal government bore indirect costs of $172 million to $776 million through USDA's Loan Deficiency Payments Program, which offers producers short- term loans and direct payments if the price of a commodity falls below the loan rate. During marketing year 2000/2001, in which StarLink was first detected in the food supply, corn prices fell below the loan rate, causing USDA to make additional income support payments to producers. In a separate study that compared the change in the price of corn with the change in the price of a substitute good, sorghum, researchers estimated that the presence of StarLink in the food supply caused a 6.8 percent drop in the price of corn, lasting for 1 year.[29] However, according to USDA, declining corn prices may have been caused by other factors as well, such as increases in supply due to favorable weather conditions or reductions in demand.

Prodigene Corn –2002

Prodigene, a biotechnology company, was responsible for two incidents in 2002 of the unauthorized release of GE corn designed to produce a protein to be used in pig vaccine, according to USDA officials. In the first incident, USDA ordered the company to destroy 155 acres of conventional corn that might have been cross-pollinated by this GE corn. In the second incident, USDA inspectors found a small number of GE corn plants growing among conventional soybeans. USDA ordered Prodigene to remove and destroy them; however, before the company did so, the soybeans were harvested and sent to a grain elevator containing 500,000 bushels of soybeans. USDA detected the problem before the soybeans were shipped from the elevator, and the agency ordered all of the soybeans destroyed. Although none of this GE corn was found in the food or feed supply, FDA issued a statement saying that the small amount of regulated material present in the soybeans would have posed no risk for human health.

Prodigene entered a consent decision with USDA in which Prodigene paid a $250,000 fine and reimbursed USDA for the destruction of the 500,000 bushels of soybeans. The company also placed $1 million in a trust fund to cover future mitigation efforts, implemented a new compliance program, and agreed to third-party audits of its field trial procedures. Despite these measures, Prodigene was involved in another incident involving GE corn in 2004. During a field trial inspection, USDA found evidence that additional GE corn may have been released to the food or feed supply. The agency ordered corrective measures and reached another settlement with Prodigene in 2007 that included a civil penalty and an agreement that neither the company nor any of its successors would apply for a GE notification or permit in the future to conduct further field trials.

Bt10 Corn – 2004

In 2004, Syngenta, a biotechnology developer, notified EPA that the company inadvertently had distributed corn seed containing an unregistered GE pesticide known as Bt10. Pesticides must be registered with EPA before commercialization. Syngenta previously determined that Bt10 was not suitable for commercialization and chose instead to register with EPA a similar pesticidal product known as Bt11. However, the company mislabeled some seed containers and, thus, inadvertently bred and sold lines of Bt10 as Bt11. Syngenta estimated that the Bt10 variety may have been planted on as many as 37,000 acres of corn, or about 1/10 of 1 percent of the annual corn acreage planted in the United States from 2001 through 2004.

In response to this incident, federal agencies took several actions. For example, although EPA determined that the protein in Bt10 was identical to the one in Bt11 and had established a tolerance exemption for Bt11, finding that there were no potential health hazards, it fined Syngenta $1.5 million for the sale of an unregistered GE pesticide.[30] FDA also concluded that the presence of Bt10 corn in the food and feed supply posed no food safety risks. In addition, USDA fined Syngenta $375,000 for moving and planting a regulated GE plant without the proper permit. In addition to the fines, Syngenta identified and destroyed all affected plants and seeds as requested by

EPA and USDA, and the company developed additional quality control mechanisms to help ensure its compliance with federal regulations.

The Bt10 incident disrupted U.S. corn exports. For example, the European Union implemented emergency inspection measures for U.S. corn from October 2005 to March 2007. In another case, South Korea required that all imports of U.S. corn be tested and certified as being free of Bt10. However, an agricultural trade group said U.S. corn exporters did not suffer a significant loss of market share due to the Bt10 incident because Syngenta paid for testing corn samples and diverting corn associated with positive samples to approved markets.

Liberty Link Rice 601 and 604 – 2006

In July 2006, another biotechnology developer, Bayer CropScience (Bayer), informed USDA that it had detected regulated genetic material in a variety of conventional long-grain rice known as Cheniere. USDA launched an investigation in August that identified the regulated material as LLRICE 601, a GE rice variety that Bayer engineered to tolerate its Liberty Link brand of herbicide. USDA investigators determined that LLRICE 601 and Cheniere had been grown at a research facility affiliated with Louisiana State University between 1999 and 2001. However, they were unable to determine conclusively that the commingling of GE and non-GE seeds, or cross-pollination took place at this facility.

Meanwhile, in response to the LLRICE 601 incident, some state and agricultural trade organizations instituted protocols for testing other rice varieties for regulated genetic material. For example, in December 2006, the Arkansas State Plant Board notified USDA that another long-grain rice variety, known as Clearfield 131 and marketed by the BASF Company, had tested positive for regulated genetic material. USDA investigators later determined that this genetic material came from another, regulated GE rice variety, LLRICE 604, also engineered by Bayer. As a result, USDA issued an emergency action notification to halt the distribution and planting of Clearfield 131. LLRICE 604 and Clearfield 131 also had been grown at the Louisiana State University research facility. However, after a year-long investigation, USDA concluded that there was insufficient information to make a conclusive link or seek an enforcement action against either Bayer or this research facility.

On November 24, 2006, USDA granted nonregulated status to LLRICE 601 on the basis of its genetic similarity to another GE rice previously approved for commercialization. However, LLRICE 604 remains regulated. In addition, FDA published statements shortly after each incident saying that the low-level presence in food or feed of the regulated genetic material from these LLRICE varieties did not pose any human health concerns. Nevertheless, despite these actions, the LLRICE incidents affected the export market for U.S. long-grain rice, which in recent years accounted for as much as 50 percent of total U.S. rice sales. Specifically, several foreign countries either banned certain varieties of U.S. rice or imposed new testing requirements on imports from the United States. For example, Japan banned the importation of U.S. long-grain rice. In another case, the European Union introduced emergency measures for the testing of U.S. rice, resulting in numerous shipments of U.S. rice being turned away from European ports. In effect, this ended rice trade between the United States and the European Union, which had accounted for as much as 10 percent of U.S. long-grain rice exports in recent years.

Furthermore, citing market disruptions caused by the LLRICE incidents, rice producers from five states filed a class action lawsuit against Bayer. As of August 2008, the plaintiffs had not yet presented estimates of rice producers' losses as a result of these incidents, but an attorney representing the plaintiffs expects the demand for total compensatory damages to be about $1 billion. These LLRICE incidents also potentially cost the BASF Company millions of dollars in lost sales of its Clearfield 131 rice. One environmental advocacy group estimated in 2007 that the worldwide costs resulting from the LLRICE incidents, including the costs associated with the loss of export markets, seed testing, elevator cleaning, and food recalls in countries where the variety of rice had not been approved, ranged from $741.0 million to $1.285 billion.[31]

Event 32 Corn – 2006

In February 2008, USDA, EPA, and FDA issued a joint public statement announcing that Dow AgroScience (Dow), a biotechnology developer, had discovered low levels of a regulated GE corn seed, called Event 32, in three lines of commercially available GE corn seed sold under the brand name Herculex. Dow engineered Event 32 to produce a pesticidal substance. According to Dow, approximately 72,000 acres were planted with corn seed containing low levels of Event 32 in 2006 and 2007. Dow's investigation of this incident concluded that the mixing of Event 32 and Herculex seed probably occurred at a single research testing field. As of August 2008, USDA's investigation was still ongoing.

Event 32 closely resembles another Dow GE corn variety, called Event 22, that is commercially available. Like Event 32, Dow engineered Event 22 to produce a pesticidal substance. Before commercialization, Event 22 was reviewed and granted nonregulated status by USDA, received a pesticide registration from EPA, and completed a food safety consultation with FDA. Given this history and the similarities between Event 32 and Event 22, the three agencies, according to USDA, affirmed that there were no public health risks posed by the low-level presence of Event 32 in food and feed. In addition, USDA and EPA concluded there were also no environmental risks. Nonetheless, USDA issued an "emergency action notification" for Event 32 seed, and EPA issued a stop-sale order. As of August 2008, these agencies were conducting investigations to determine whether any violations had occurred. According to Dow, it voluntarily recalled unplanted seed containing Event 32. Dow also provided USDA with the testing method it used to detect Event 32. However, USDA said this test may not be sensitive enough to detect the low levels of Event 32 expected in the commercial seed supply.

The Event 32 incident did not lead to detectable economic impacts. To preclude trade disruptions, USDA provided relevant information to U.S. trading partners, including information on the similarities between Event 32 and Event 22, noting that the latter GE variety is accepted by a number of countries, including Japan, the largest purchaser of U.S. corn.

Appendix VIII: GE Regulatory Issues and Alternatives Discussed in USDA's DEIS and Proposed Rule

USDA's DEIS, announced in the *Federal Register* on July 17, 2007, presents various issues and alternatives for regulating GE organisms, including crops. Table 6 summarizes these issues and alternatives; alternatives in bold type indicate USDA's preliminary preferred options in the DEIS. USDA invited public comments on these issues and alternatives by September 11, 2007. On October 9, 2008, after considering the comments on the DEIS and other factors, USDA published a proposed rule that, if adopted, would amend its regulations for GE organisms, including plants. According to USDA, differences between the proposed rule and the DEIS are primarily a matter of reorganizing and realigning some materials and their corresponding regulatory alternatives, using more descriptive terms in some criteria listed in the alternatives, and choosing between regulatory alternatives that fall within the analysis of the DEIS. The proposed rule contains a table that provides a comparison between the proposed changes in the rule and DEIS. Specifically, it indicates which of the DEIS alternatives most closely match the proposed rule. We have included that information in table 6.

Table 6. GE Regulatory Issues and Alternatives Discussed in USDA's DEIS and Proposed Rule

Issue	Alternatives considered (USDA's preliminary preference is shown in bold type)	Alternative(s) in DEIS that correspond to proposed change(s) to regulations
1 – Broadening Regulatory Scope to Include GE Crops Posing Noxious Weed Risk USDA is considering the broadening of its regulatory scope beyond GE organisms that may pose a plant pest risk to include GE crops that may pose a noxious weed risk and GE organisms that may be used to control noxious weeds or plant pests (biological control agents). Do regulatory requirements for these organisms need to be established?	1. No action—continue to regulate GE organisms as potential plant pests, and use genetic transformation as the trigger for regulation (event by event). **2. Expand the scope of what is regulated by adding considerations of noxious weed risk and regulating GE biological control organisms in addition to evaluating plant pest risks, and use genetic transformation as the trigger for regulation. Continue to regulate event by event.** 3. Expand the scope of what is regulated by adding considerations of noxious weed risk and regulating GE biological control organisms in addition to evaluating plant pest risks. Use novelty of the trait in the species as the trigger for regulation. **4. Exclude specific classes of highly familiar organisms and highly domesticated, nonweedy crop plants and, potentially, those regulated by another federal agency from regulation.** *USDA's explanation*: The second alternative would eliminate potential gaps that may occur as genetic engineering techniques continue to advance. The fourth	2 or 3[a]

Table 6. (Continued)

Issue	Alternatives considered (USDA's preliminary preference is shown in bold type)	Alternative(s) in DEIS that correspond to proposed change(s) to regulations
	alternative would allow USDA and a developer to focus resources on GE crops that have a higher potential risk.	
2 – Use of Risk-Based Categories for New Products USDA is considering revisions to the regulations to increase transparency and to address advances in technology that may create new products and concerns. Should a new system of risk-based categories be designed to deal with new products and new concerns? If so, what criteria should be used to establish the risk-based categories?	1. No action—continue to use a two-tiered system (notifications and permits). 2. Abolish categories and treat all future proposals for the introduction of GE organisms on a case-by-case basis. 3. Establish a tiered permitting system for all organisms based on newly devised criteria. **4. Establish a tiered permitting system for plants based on newly devised criteria and evaluate permit applications for introductions of nonplant organisms on a case-by-case basis.** *USDA's explanation*: The fourth alternative would be more transparent, allowing developers and the public to see that organisms are to be regulated on the basis of risk and familiarity.	4
3 – Regulatory Flexibility to Allow Commercialization Despite Minor Unresolved Risks	1. No action—continue with current system granting full nonregulated status to crops that removes them from all regulatory obligations.	2

Table 6. (Continued)

Issue	Alternatives considered (USDA's preliminary preference is shown in bold type)	Alternative(s) in DEIS that correspond to proposed change(s) to regulations
USDA is considering ways to provide regulatory flexibility for future decisions by accommodating commercialization of certain GE organisms while continuing, in some cases, to regulate the organisms on the basis of minor unresolved risks. Other regulated articles could be treated as they have been under the current system, in which all regulatory restrictions are removed. What environmental factors should be considered in distinguishing between these kinds of decisions?	**2. Continue to allow for the option of granting full nonregulated status and develop appropriate criteria and procedures through which crops can be removed from permitting, but some degree of agency oversight, as necessary, to mitigate any minor risks is retained.** *USDA's explanation*: Under the second alternative, the added flexibility of being able to retain some oversight may be useful for some types of GE organisms that might be developed in the future.	
4 – Regulation of Crops Producing Pharmaceutical and Industrial Compounds Are there changes that should be considered relative to environmental	1. No action—continue to allow food and feed crops to be used for the production of pharmaceutical and industrial compounds and to allow field testing under very stringent conditions. **2. Continue to allow food and feed crops to be used for the production of pharmaceutical and industrial**	2

Table 6. (Continued)

Issue	Alternatives considered (USDA's preliminary preference is shown in bold type)	Alternative(s) in DEIS that correspond to proposed change(s) to regulations
review of, and permit conditions for, GE crops that produce pharmaceutical and industrial compounds?	**compounds. The agency would impose confinement requirements, as appropriate, based on the risk posed by the organism and would consider food safety in setting conditions.** 3. Do not allow crops producing substances not intended for food uses to be field tested, that is, these crops could be grown only in contained facilities. 4. Allow field testing only if the crop has no food or feed uses. 5. Allow field testing of food/feed crops producing substances not intended for food uses only if food safety has been addressed. *USDA's explanation*: Under the second alternative, the use of highly stringent confinement measures can be used to protect the environment from significant impact and the consideration of food safety will further enhance human safety.	
5 – Regulation of Nonviable Plant Material The definition of noxious weeds in the Plant Protection Act includes not only plants, but also plant products. On the	1. No action—do not regulate nonviable GE material. **2. Regulate nonviable GE plant material in certain circumstances, on the basis of the risks posed.** 3. Regulate all nonviable GE plant material.	2

Table 6. (Continued)

Issue	Alternatives considered (USDA's preliminary preference is shown in bold type)	Alternative(s) in DEIS that correspond to proposed change(s) to regulations
basis of that authority, USDA is considering the regulation of nonviable plant material (i.e., plant materials, such as stems and leaves, that do not propagate new plants). Is the regulation of nonviable material appropriate and, if so, in which cases should we regulate?	*USDA's explanation*: The second alternative is preferred because, in most cases, nonviable plant material will not pose a risk. However, in some cases, oversight might be required to ensure the safe handling and disposal of this material.	
6 – New Mechanism for Regulating Nonfood/Nonfeed Crops Producing Pharmaceutical and Industrial Compounds USDA is considering establishing a new mechanism involving USDA, the states, and the producer for commercial production of plants not intended for food or feed in cases where the produ-er would prefer to develop and extract pharmaceutical and industrial compou-ds under confinement conditions with governmental oversight, rather than USDA granting nonregulated status to	1. No action—continue to authorize field tests of crops not intended for food or feed use under permit. Require application and review of these permits on an annual basis. **2. Allow for special multiyear permits, with ongoing oversight. The new system would maintain these crops under regulation, but USDA oversight would be exercised in a different manner than under the current system of permits.** *USDA's explanation*: Under the second alternative, the new system would be just as protective of the enviro-nment as the current system, but in a manner that is more efficient.	1[b]

Table 6. (Continued)

Issue	Alternatives considered (USDA's preliminary preference is shown in bold type)	Alternative(s) in DEIS that correspond to proposed change(s) to regulations
these plants. What should be the haracteristics of this mechanism?		
7 – Allowance for Low-Level Presence of Regulated GE Material in Crops, Food, Feed, or Seed The current regulations have no provision for the low-level presence of regulated articles in commercial crops, food, feed, or seed of GE plant material that has not completed the required regulatory processes. Should low-level occurrences of a regulated article be exempted from regulation?	1. No action—allow field testing to continue using current confinement strategies to reduce the likelihood of regulated articles occurring in commercial commodities or seeds. 2. Establish criteria under which occurrence of regulated articles would be allowable, that is, considered not-actionable by USDA. Do not allow field testing of crops that do not meet all of these criteria, including addressing food safety issues if applicable (i.e., if the GE plant is a food crop). **3. Establish criteria under which occurrence of regulated articles would be allowable, that is, considered not-actionable by USDA. Allow field testing and impose confinement strategies based on whether a plant meets the criteria.** 4. Impose a very strict confinement regime on all field tests, as is currently done for pharmaceutical and industrial crops, that would further reduce the likelihood of regulated articles occurring in commercial	3

Table 6. (Continued)

Issue	Alternatives considered (USDA's preliminary preference is shown in bold type)	Alternative(s) in DEIS that correspond to proposed change(s) to regulations
	commodities or seeds. *USDA's explanation*: The agency's analysis indicates that material meeting the safety-based criteria of the third alternative would not pose a risk for significant environmental impact.	
8 – Risk Assessment for Imported GE Commodities Should USDA provide expedited review or exemption from review for certain low-risk, imported GE commodities intended for food, feed, or processing that have received all necessary regulatory approvals in their country-of-origin and are not intended for propagation in the United States?	1. No action—continue to evaluate commodity importation requests on a case-by-case basis. **2. Establish criteria that will be applied to determine the appropriate level of risk assessment for imported GE commodities. This alternative could include a decision to exempt certain organisms or to allow importation under conditions that minimize environmental release.** 3. Disallow importation of any commodity pending full USDA approval for deregulation. 4. Accept any importation of a product from a foreign country that has evaluated the safety of the product and approved it for unconfined environmental release. 5. Accept any importation of a product from a foreign country that has evaluated the safety of the product and approved it for unconfined environmental release using a review process equivalent to USDA's.	1[c]

Table 6. (Continued)

Issue	Alternatives considered (USDA's preliminary preference is shown in bold type)	Alternative(s) in DEIS that correspond to proposed change(s) to regulations
	USDA's explanation: Under the second alternative, the proposed exemption criteria should ensure that exempted GE commodities would not result in significant environmental impacts, even if an environmental release should accidentally occur.	
9 – Interstate Movement of Well-Studied, Low-Risk GE Material Currently, GE *Arabidopsis* (a mustard plant commonly used in genetics research) is exempt from interstate movement restrictions because they are well-understood and extensively used in research. Should the movement of GE *Arabidopsis* or other GE organisms be exempted from movement restriction?	1. No action—require interstate movement authorizations for all organisms on the list in current regulations. **2. Exempt a class of GE crops or organisms that are well-studied and present little or no environmental risk from permit requirements for interstate movement as is currently done for *Arabidopsis*.** 3. Create a process to apply for an interstate movement exemption for a particular species. *USDA's explanation:* Regarding the second alternative, an expansion of the exempted list to include other well-studied research organisms would present little or no risk of significant environmental impact.	3[d]
10 – Container Requirements for Shipping GE Material	1. No action—retain current list of approved containers and issue variances when necessary.	2

Table 6. (Continued)

Issue	Alternatives considered (USDA's preliminary preference is shown in bold type)	Alternative(s) in DEIS that correspond to proposed change(s) to regulations
What environmental considerations should be evaluated if USDA were to move from prescriptive container requirements for shipment of GE organisms to performance- based container requirements, supplemented with guidance on ways to meet the performance standards?	**2. Switch to performance-based standards for all transport containers.** 3. Expand current list of approved containers and issue variances when necessary. *USDA's explanation:* Under the second alternative, having performance-based standards would eliminate the need for variances, reduce the burden on applicants, and increase the efficient use of agency resources while protecting the environment.	

Source: USDA's DEIS, "Introduction of Genetically Engineered Organisms." The DEIS's availability for review was announced in the *Federal Register* on July 17, 2007. *(72 Fed. Reg.* 39,021)

[a]According to the proposed rule, USDA would regulate GE plants either on the basis that (1) the parent plant from which the GE plant was derived is a plant pest or noxious weed, (2) the trait introduced by genetic engineering could increase the potential of the GE plant to be a plant pest or noxious weed, (3) the risk that the GE plant poses as a plant pest or noxious weed is unknown, or (4) the Administrator of APHIS determines that the GE plant poses a plant pest or noxious weed risk. As such, aspects of both DEIS alternatives 2 and 3 are incorporated into the proposed rule.

[b]According to the proposed rule, USDA concluded that the current permitting procedures and the use of stringent permitting conditions would effectively minimize the risk associated with the environmental release of pharmaceutical or industrial compounds.

[c]USDA stated in the proposed rule that it is not proposing criteria to evaluate risks of GE imported commodities that would allow it to conduct expedited reviews, but it does not rule out the possibility of developing such a system in the future.

[d]According to the proposed rule, USDA would retain existing conditional exemptions from permitting requirements for the interstate movement of certain GE organisms but is not proposing new exemptions. Instead, the agency is proposing a petition process for approving additional exemptions.

End Notes

[1]GAO, *Results-Oriented Government: Practices That Can Help Sustain Collaboration among Federal Agencies*, GAO-06-15 (Washington, D.C.: Oct. 21, 2005).

[2] Not all field trials authorized under USDA permits or notifications are carried out. A GE crop developer may decide not to plant the field trial if, for example, the seeds have not performed as expected in laboratory testing, the necessary quantity of seeds is not available, or the weather is not favorable.

[3] As of July 10, 2008, USDA had not resolved 32 additional incidents. IES is located within USDA's Animal and Plant Health Inspection Service.

[4] A tolerance from EPA establishes the maximum amount of pesticidal residue allowed on food or feed.

[5] The guidance does not apply to GE pesticides, which are regulated by EPA. Nor does the guidance apply to plants used to produce pharmaceutical compounds.

[6] See GAO-06-15. GAO also identified an eighth practice—that is, reinforcing individual accountability for collaborative efforts through performance management systems—that we do not address in this chapterbecause it was beyond the scope of our work.

[7] FDA has subjected only one substance added to a GE crop—a protein added to a tomato engineered for delayed ripening—to its food additive review process, and this was done at the request of the developer.

[8] The National Economic Council is a part of the White House's Office of Policy Development. The council advises the President on matters related to U.S. and global economic policy.

[9] Codex Alimentarius sets international food safety standards.

[10] The council has published three relevant reports at the request of USDA: *Ecological Monitoring of Genetically Modified Crops: A Workshop Summary (2001); Environmental Effects of Transgenic Plants: The Scope and Adequacy of Regulations* (2002); and *Biological Confinement of Genetically Engineered Organisms* (2004). The National Research Council is part of the National Academy of Science—a private, nonprofit organization comprising distinguished scientists and engineers with a mandate from Congress requiring it to advise the federal government on scientific and technical matters.

[11] *Environmental Effects of Transgenic Plants.*

[12] The council is a cabinet-level organization that includes representatives from USDA, EPA, FDA, and other federal agencies.

[13] National Science and Technology Council, *Agricultural Biotechnology Risk Analysis Research in the Federal Government: Cross Agency Cooperation* (2006).

[14] Planting a "refuge" of crops that do not contain the pesticide Bt near crops that do contain Bt is intended to reduce the likelihood that insect populations will develop a resistance to Bt.

[15] Advisory Committee on Biotechnology and 21st Century Agriculture, *What Issues Should USDA Consider Regarding Coexistence among Diverse Agricultural Systems in a Dynamic, Evolving, and Complex Marketplace?* (March 2008).

[16] GAO, *Genetically Modified Foods: Experts View Regimen of Safety Tests as Adequate, but FDA's Evaluation Process Could Be Enhanced*, GAO-02-566 (Washington, D.C.: May 23, 2002).

[17] Under the Administrative Procedure Act, agencies generally provide "interested persons" with an opportunity to comment on proposed rules, and agencies generally respond to the issues and matters raised in those comments in their final rules. The comments we analyzed are from stakeholders who chose to submit comments to USDA; therefore, they are not necessarily representative of all stakeholders who might have insights or opinions regarding biotechnology regulation.

[18] The deadline for public comments on the proposed rule is November 24, 2008. After considering these comments, the agency plans to issue a final rule accompanied by a Final Environmental Impact Statement. However, the dates for these publications are uncertain.

[19] In its fiscal fear 2009 budget request, the administration is requesting a 38 percent increase in funding—from $11.7 million to $16.2 million—and a 28 percent increase in staffing—from 74 staff years to 95 staff years—for USDA's Biotechnology Regulatory Services.

[20] Virus coat protein GE pesticides are derived from the genetic material that plant viruses commonly use for protection.

[21] GAO, *Results-Oriented Government: Practices That Can Help Sustain Collaboration among Federal Agencies*, GAO-06-15 (Washington, D.C.: Oct. 21, 2005).

[22] 51 *Fed. Reg.* 23,302 (June 26, 1986). The announcement followed publication of a proposed coordinated framework in 1984. 49 *Fed. Reg.* 50,856 (Dec. 31, 1984).

[23] Because states have primary responsibility for pesticide use within their borders, once a pesticide is registered with EPA, the producer may also be required to register the pesticide with state authorities. State registration may involve more stringent requirements on how the pesticide is used.

[24] See 21 U.S.C. § 331. A food is deemed adulterated if, among other things, it contains any added poisonous or deleterious substance that may render the food injurious to health or if it contains an unapproved food additive. See 21 U.S.C. § 342.

[25] FDA established its basic policy regarding the review of GE foods in its 1992 S*tatement of Policy: Foods Derived from New Plant Varieties*. In 1997, FDA supplemented its 1992 policy with the current *Guidance on Consultation Procedures*, clarifying procedures for the initial and final consultations.

[26] According to FDA, 10 Investigational New Drug applications for pharmaceutical products derived from GE crops have been submitted. As of July 2008, only two of the applications were active, and neither involved the use of food plants.

[27] In December 2000, an EPA science advisory panel concluded that the pesticidal protein in StarLink had a medium probability of being a potential allergen. However, the Centers for Disease Control and Prevention, in a 2001 study of this protein's allergenicity conducted for FDA, reported that "although the study participants may have experienced allergic reactions, based upon the results of this study alone, we cannot confirm that a reported illness was a food-associated allergic reaction."

[28] T. Schmitz, A. Schmitz, and C. Moss, "The Economic Impact of Starlink Corn," *Agribusiness,* vol. 21, no. 3 (2005).

[29] C. Carter and A. Smith, "Estimating the Market Effect of a Food Scare: The Case of Genetically Modified StarLink Corn," *Review of Economics & Statistics*, vol. 89, no. 3 (2007).

[30] The initial penalty exceeded $6 million, but Syngenta qualified for a 75 percent reduction due to mitigating circumstances, including its voluntary disclosure of the incident and cooperation with EPA during the subsequent investigation.

[31] Neal E. Blue, *Risky Business: Economic and Regulatory Impacts from the Unintended Release of Genetically Engineered Rice Varieties into the Rice Merchandising System of the U.S.*, Greenpeace International (November 2007).

In: Genetically Engineered Crops and Proposed... ISBN: 978-1-60876-067-1
Editors: Brandon C. Vaughn

Chapter 2

THE FIRST DECADE OF GENETICALLY ENGINEERED CROPS IN THE UNITED STATES

Jorge Fernandez-Cornejo and Margriet Caswell
U.S. Department of Agriculture (USDA)

ABSTRACT

Ten years after the first generation of genetically engineered (GE) varieties became commercially available, adoption of these varieties by U.S. farmers is widespread for major crops. Driven by farmers' expectations of higher yields, savings in management time, and lower pesticide costs, the adoption of corn, soybean, and cotton GE varieties has increased rapidly. Despite the benefits, however, environmental and consumer concerns may have limited acceptance of GE crops, particularly in Europe. This chapterfocuses on GE crops and their adoption in the United States over the past 10 years. It examines the three major stakeholders of agricultural biotechnology and finds that (1) the pace of R&D activity by producers of GE seed (the seed firms and technology providers) has been rapid, (2) farmers have adopted some GE varieties widely and at a rapid rate and benefited from such adoption, and (3) the level of consumer concerns about foods that contain GE ingredients varies by country, with European consumers being most concerned.

Keywords: genetically engineered crops, agricultural biotechnology, seed industry, research and development, adoption, crop yields, pesticide use, corn, soybeans, cotton

Acknowledgments

The authors wish to thank William Lin, Paul Heisey, Keith Wiebe, Marca Weinberg, Utpal Vasavada, and Mary Bohman of ERS for the helpful comments provided on earlier drafts of this report. We also thank Andrew Rude from the Foreign Agricultural Service, Neil E. Hoffman from the Animal and Plant Health Inspection Service, John W. Radin from the Agricultural Research Service, Daniel Jones from the Cooperative State Research, Education, and Extension Service, and Matthew C. Rousu from Susquehanna University. We also thank Lou King for editorial assistance and Anne Pearl for graphics and layout.

Summary

Over the past decade, developments in modern biotechnology have expanded the scope of biological innovations by providing new tools for increasing crop yields and agricultural productivity. The role that biotechnology will play in agriculture in the United States and globally will depend on a number of factors and uncertainties. What seems certain, however, is that the ultimate contribution of agricultural biotechnology will depend on our ability to identify and measure its potential benefits and risks.

What Is the Issue?

Ten years after the first generation of genetically engineered (GE) varieties of major crops became commercially available, adoption of these varieties by U.S. farmers has become widespread. United States consumers eat many products derived from these crops—including some cornmeal, oils, sugars, and other food products—largely unaware of their GE content. Despite the rapid increase in the adoption of GE corn, soybean, and cotton varieties by U.S. farmers, questions remain regarding the impact of agricultural biotechnology. These issues range from the economic and environmental impacts to consumer acceptance.

What Did the Study Find?

This study examined the three major stakeholders in agricultural biotechnology: **seed suppliers and technology providers, farmers, and consumers.**

Seed suppliers/technology providers. Strengthening of intellectual property rights protection in the 1970s and 1980s increased returns to research and offered greater incentives for private companies to invest in seed development and crop biotechnology. Since 1987, seed producers have submitted nearly 11,600 applications to USDA's Animal and Plant Health Inspection Service for field testing of GE varieties. More than 10,700 (92 percent) have been approved. Approvals peaked in 2002 with 1,190. Most approved applications involved major crops, with nearly 5,000 for corn alone, followed by soybeans, potatoes, and cotton. More than 6,600 of the approved applications included GE varieties with herbicide tolerance or insect resistance. Significant numbers of applications were approved for varieties with improved product quality, viral resistance, and enhanced agronomic properties such as drought and fungal resistance.

Farmers. Adoption of GE soybeans, corn, and cotton by U.S. farmers has increased most years since these varieties became commercially available in 1996. By 2005, herbicide-tolerant soybeans accounted for 87 percent of total U.S. soybean acreage, while herbicide-tolerant cotton accounted for about 60 percent of total cotton acreage. Adoption of insect-resistant crops is concentrated in areas with high levels of pest infestation and varies across States. Insect-resistant cotton was planted on 52 percent of cotton acreage in 2005—ranging from 13 percent in California to 85 percent in Louisiana. Insect-resistant corn accounted for 35 percent of the total acreage in 2005, following the introduction of a new variety to control the corn rootworm.

The economic impact of GE crops on producers varies by crop and technology. Herbicide-tolerant cotton and corn were associated with increased returns, as were insect-resistant cotton and corn when pest infestations were more prevalent. Despite the rapid adoption of herbicide-tolerant soybeans, there was little impact on net farm returns in 1997 and 1998. However, the adoption of herbicide-tolerant soybeans is associated with increased off- farm household income, suggesting that farmers adopt this technology because the simplicity and flexibility of the technology permit them to save management time, allowing them to benefit from additional income from off-farm activities.

Genetically engineered crops also seem to have environmental benefits. Overall pesticide use is lower for adopters of GE crops, and the adoption of herbicide-tolerant soybeans may indirectly benefit the environment by encouraging the adoption of soil conservation practices.

Consumers. Most surveys and consumer studies indicate consumers have at least some concerns about foods containing GE ingredients, but these concerns have not had a large impact on the market for these foods in the United States. Despite the concerns of U.S. consumers, "GE-free" labels on foods are not widely used in the United States. Manufacturers have been active in creating a market for GE-free foods. Between 2000 and 2004, manufacturers introduced more than 3,500 products that had explicit non- GE labeling, most of them food products.

In the European Union and some other countries, however, consumer concerns have spurred a movement away from foods with GE ingredients. Despite the fact that some European consumers are willing to consume foods containing GE ingredients, very few of these foods are found on European grocery shelves.

How Was the Study Conducted?

This chapterexamined the three major stakeholders of agricultural biotechnology: GE seed suppliers and technology providers (biotech firms), farmers, and consumers. To examine biotech and seed firms, we used information from the literature as well as from the database of USDA approvals of field testing for new GE varieties. To study seed users, we drew on ERS studies based on USDA farm surveys, and to review the consumer perspective, we summarized surveys of consumers' attitudes from the literature.

Introduction

Over the past decade, developments in modern biotechnology have expanded the scope of biological innovations by providing new tools for increasing crop yields and agricultural productivity. Agricultural biotechnology is a collection of scientific techniques, including genetic engineering, that are used to create, improve, or modify plants, animals, and microorganisms. Genetic engineering (GE) techniques allow a precise alteration of a plant's traits (facilitating the development of characteristics not possible through traditional plant breeding), and permit targeting of a single plant trait (decreasing the number of unintended characteristics that may occur with traditional breeding).[1]

The commercial success of GE crop varieties typically requires that biotechnology-derived trait enhancements be incorporated into successful cultivars (cultivated varieties with useful agronomic properties), the development of which requires significant knowledge of traditional plant breeding and the availability of genetic material (germplasm). This complementarity has been related to various institutional arrangements between seed and technology suppliers.

GE crops are often classified into one of three generations (Panos). Crops with enhanced input traits, such as herbicide tolerance, insect resistance, and tolerance to environmental stresses (like drought), represent the first generation. GE crops benefit farmers and may also offer environmental benefits. Second-generation crops include those with added-value output traits, such as nutrient enhancement for animal feed. Consumers will benefit directly from these products when they are available on the market. The third generation includes crops that produce pharmaceuticals or improve processing of bio-based fuels, and products beyond traditional food and fiber. At present, adoption of GE crops is generally limited to those with first-generation traits, which were tested on a large scale (field testing) in the 1980s to ensure that the desired traits will perform under production conditions. Second- and third-generation GE crops are in various stages of research and development.

Ten years after the first generation of GE varieties became commercially available, they have been widely adopted by U.S. farmers, driven by expectations of higher yields, savings in management time, and lower pesticide costs. Despite these benefits, environmental and consumer concerns may have limited acceptance of agricultural biotechnology, particularly in Europe. In the United States, foods containing GE ingredients currently available in the U.S. market do not require labels, since the U.S. Food and Drug Administration has determined that these foods are "substantially equivalent" to their non-GE counterparts (Shoemaker et al., 2003; FDA, 1992). Thus, U.S. consumers have been eating foods that contain GE ingredients (corn meal, oils, sugars) for the past 10 years while remaining largely unaware of their GE content.

Rapid Change and Pace of R&D Activity Characterize the Seed Industry and Technology Providers

The U.S. commercial seed market is the world's largest—with an estimated annual value of $5.7 billion per year in the late 1990s—followed by China at $3 billion and Japan at $2.5 billion (Fernandez-Cornejo, 2004). Moreover, the U.S. seed market is growing (in quantity and value), mainly because farmers have been increasing purchases of seed and reducing the planting of saved seed. Growth in the seed market has been particularly rapid for major field crops—corn, soybeans, cotton, and wheat—that together constituted two-thirds of the seed market value in 1997.

The U.S. seed industry began a transformation in the 1930s, with the introduction of commercially viable hybrid seeds. These hybrids were higher yielding than nonhybrid varieties but degenerative, so farmers had to purchase new seed every year to maintain the high yields. Further changes were motivated by the strengthening of intellectual property rights (IPR) protection, mainly during the 1970s and 1980s, which increased returns to research and offered a greater incentive for private companies to invest in seed development. The two principal forms of legal protection are plant variety protection (PVP) certificates issued by the Plant Variety Protection Office of USDA and patents issued by the U.S. Patent and Trademark Office. Both grant private crop breeders exclusive rights to multiply and market their newly developed varieties. However, patents provide more control since PVP certificates have a research exemption allowing others to use the new variety for research purposes. Agricultural biotechnology patents, mostly dealing with some aspect of plant breeding, have outpaced the general upward trend in patenting throughout the U.S. economy. During 1996-2000, 75 percent of over 4,200 new agricultural biotech patents went to private industry (King and Heisey).

Enhanced protection of intellectual property rights brought rapid increases in private research and development (R&D) investments and changes in market concentration in the U.S. seed industry. R&D expenditures on plant breeding for many major crops shifted from mainly public to mainly private. Private spending on crop variety R&D increased fourteenfold between 1960 and 1996 (adjusted for inflation), while public expenditures changed little (Fernandez-Cornejo, 2004).

As the amount of private capital devoted to R&D in the seed industry grew rapidly, the number of private firms engaged in plant breeding also grew, until peaking in the early 1990s. Subsequently, the seed industry consolidated, with fewer firms capable of sustaining the research investment needed to develop new seed varieties. Mergers and acquisitions created a seed industry structure dominated by large companies with primary investments in related sectors, such as pharmaceutical, petrochemical, and food (Fernandez-Cornejo, 2004).[2]

In the early 1980s, developments in biological sciences created an additional incentive for private firms to increase their investment in R&D and seed production. As the first products of crop biotechnology were tested on a large scale in the 1980s, the seed industry's structure underwent additional transformation. Companies sought to achieve economies of scale to offset the high costs of biotechnology R&D through an extensive process of mergers, acquisitions, and joint ventures. Chemical and seed businesses combined to take advantage of strong demand complementarities between products (Just and Hueth, 1993). For example, the herbicide glyphosate and soybean seeds tolerant to glyphosate are sold by the same firm. As a consequence of the merger activity, the seed industry became more concentrated. By 1997, the share of U.S. seed sales (including GE and conventional varieties) controlled by the four largest firms providing seed of each crop reached 92 percent for cotton, 69 percent for corn, and 47 percent for soybeans (table 1).

From the Laboratory to the Field

A critical part of new variety development is field-testing to ensure that the desired traits will perform under production conditions. The release of new GE varieties of organisms into the environment is regulated through field release permits and monitored by USDA's Animal and Plant Health Inspection Service (APHIS) (see box, "Regulatory Oversight"). The number of field releases of plant varieties for testing purposes provides a useful indicator of R&D efforts on crop biotechnology.

By early April 2005, nearly 11,600 applications had been received by APHIS since 1987 and more than 10,700 (92 percent) had been approved (Virginia Polytechnic Institute and State University, 2005). Approvals peaked in 2002 with 1,190 (figure 1). Most applications approved for field testing involved major crops, particularly corn with nearly 5,000 applications approved, followed by soybeans, potatoes, cotton, tomatoes, and wheat (figure

2). Applications approved between 1987 and early April 2005 included GE varieties with herbicide tolerance (3,587), insect resistance (3,141), improved product quality (flavor, appearance, or nutrition) (2,314), virus resistance (1,239), and agronomic properties like drought resistance (1,043) and fungal resistance (647) (figure 3).

REGULATORY OVERSIGHT

Before commercial introduction, genetically engineered crops must conform to standards set by State and Federal statutes (USDA, 2005). Under the Coordinated Framework for the Regulation of Biotechnology, Federal oversight is shared by the U.S. Department of Agriculture (USDA), the U.S. Environmental Protection Agency (EPA), and the U.S. Food and Drug Administration (FDA).

USDA's Animal and Plant Health Inspection Service (APHIS) plays a central role in regulating field-testing of agricultural biotechnology products. Through either a notification or permit procedure, such products, which include genetically engineered plants, microorganisms, and invertebrates, are considered "regulated articles." APHIS determines whether to authorize the test, based on whether the release will pose a risk to agriculture or the environment. After years of field tests, an applicant may petition APHIS for a determination of nonregulated status in order to facilitate commercialization of the product. If, after extensive review, APHIS determines that the unconfined release does not pose a significant risk to agriculture or the environment, the organism is "deregulated." At this point, the organism is no longer considered a regulated article and can be moved and planted without APHIS authorization (USDA, 2004).

If a plant is engineered to produce a substance that "prevents, destroys, repels, or mitigates a pest," it is considered a pesticide and is subject to regulation by EPA (*Federal Register*, November 23, 1994). FDA regulates all food applications of crops, including those crops that are developed through the use of biotechnology, to ensure that foods derived from new plant varieties are safe to eat. A more complete description of the EPA and FDA regulations of GE products may be found in EPA (2003) and FDA (1992, 2005).

Though the current regulatory system is considered to be effective, USDA, EPA, and FDA continuously look forward and make necessary changes to address new trends and issues of the future. For example,

USDA's APHIS has made updates in 1993 and 1997 and is currently considering the need for additional changes in the regulations (USDA, 2004). The National Academy of Sciences also issued a report that made recommendations suggesting that regulation "could be improved further" by making the process more "transparent and rigorous" by enhanced scientific peer review, solicitation of public input, and "more explicit presentation of data, methods, analyses, and interpretations" (NRC, 2003).

Table 1. Estimated U.S. seed market shares for major field crops, 1997

Company	Corn	Soybean	Cotton
Pioneer Hi-Bred	42.0	19.0	
Monsanto[1]	14.0	19.0	11.0
Novartis	9.0	5.0	
Delta & Pine Land[2]			73.0
Dow Agrosciences/Mycogen	4.0	4.0	
California Planting Seed Distributors			6.0
All-Tex			2.0
Four largest total	69.0	47.0	92.0

[1] Monsanto acquired DeKalb in 1997 and Asgrow in 1998.

[2] The merger proposed between Monsanto and Delta & Pine Land in 1998 was called off in December 1999.

Source: Fernandez-Cornejo, 2004.

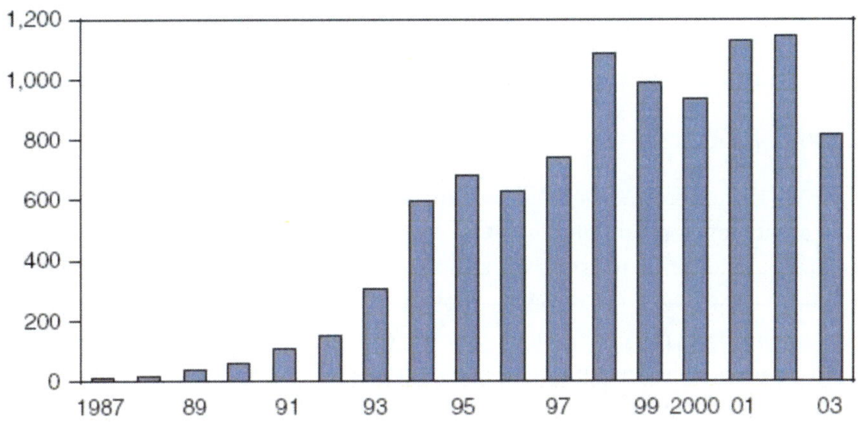

Source: Virginia Polytechnic Institute and State University, 2005.

Figure 1. Permits for release of GE varieties approved by APHIS

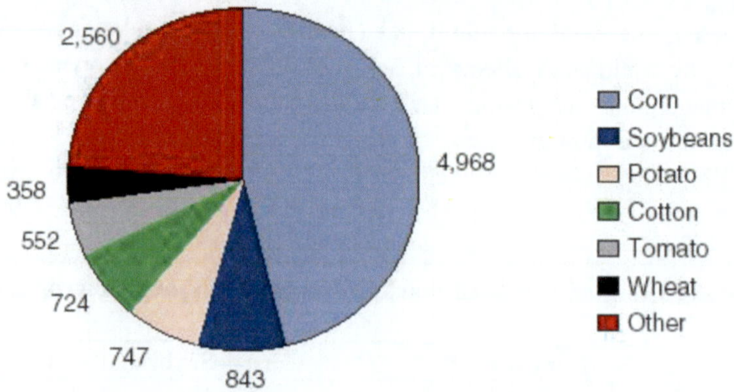

Source: Virginia Polytechnic Institute and State University, 2005.

Figure 2. Total number of permits approved by APHIS, by crop

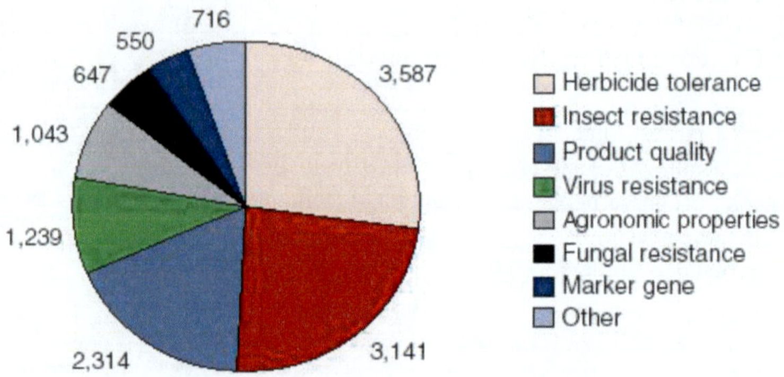

Source: Virginia Polytechnic Institute and State University, 2005.

Figure 3. Total number of permits approved by APHIS, by GE trait

APHIS approvals for field testing also provide an indication of products that are in development and that may come "through the pipeline" in the future (table 2). In addition to crops with improved pest management traits, approvals include crops with traits that provide viral/fungal resistance, favorable agronomic properties (resistance to cold, drought, salinity, more efficient use of nitrogen), enhanced product quality (delayed ripening, increased protein and oil content, modified starch content, nutraceuticals (added vitamins, iron, antioxidants such as beta-carotene), and pharmaceuticals. Additional information may be found in Runge and Ryan and in Pew Initiative on Food and Biotechnology (2001).

Table 2. Biotech crops currently available and in development in the United States

Crop	Input traits				Output traits	
	Herbicide tolerance	**Insect resistance**	**Viral/fungal resistance**	**Agronomic properties**[9]	**Product quality**[11]	**Nutraceuticals; pharmaceuticals; industrial**[13]
Corn	C	C[5]	D	D	D	D
Soybeans	C	D		D	D	
Cotton	C	C[6]		D	D	
Potatoes		W[7]	D	D	D	D
Wheat	C[2]		D			
Other field crops[1]	C[3] D[4]	D	D	D	D	D
Tomato, sq-uash, melon			D	D	W[12] D	D
Other vegetables	D				D	
Papaya			C[8]			
Fruit trees			D		D	
Other trees				D[10]	D	
Flowers					D	

C = Currently available; D = In various stages of development and testing; W = Withdrawn from the market.

Sources: Virginia Polytechnic Institute and State University; USDA, APHIS; Colorado State University; Shoemaker et al.; Pew.

[1]Includes barley, canola, peanuts, tobacco, rice, alfalfa, etc.

[2]Monsanto discontinued breeding and field-level research on its GE Roundup Ready wheat in 2004.

[3]Canola.

[4]Barley, rice, sugar beets.

[5]Bt corn to control the corn borer commercially available since 1996; Bt corn for corn rootworm control commercially available since 2003.

[6]Bt cotton to control the tobacco budworm, the bollworm, and the pink bollworm, commercially available since 1996.

[7]Bt potatoes, containing built-in resistance to the Colorado potato beetle, were commercially introduced in 1996 and withdrawn in 1999.

[8]In the mid 1990s, researchers at Cornell University and at the University of Hawaii developed two virus-resistant varieties of GE papaya. First commercial plantings were made in 1998. The new varieties were proved successful in resisting a viral epidemic and were planted on more than 30 percent of Hawaii's papaya acreage in 1999.

[9]Resistance to cold, drought, frost, salinity; more efficient use of nitrogen; increased yield.

[10]Modified lignin content (for example, to reduce cost of paper making from trees).

[11]Includes delayed ripening; increased protein, carbohydrate, fatty acid, micronutrient, oil, and modified starch content; enhanced flavor and texture (fruits and vegetables); color (cotton, flowers); fiber properties (cotton); gluten content; natural decaffeination; and low phytase.

[12]Tomato genetically engineered to remain on the vine longer and ripen to full flavor after harvest; currently withdrawn from the market (Colorado State University, 2004).

[13]Includes increased vitamin, iron, beta-carotene content; antibodies, vaccines; specialty machine oils.

After extensively field-testing a GE variety, an applicant may petition APHIS to deregulate (grant nonregulated status) the variety. If, after extensive review, APHIS determines that the new variety poses no significant risk to agriculture or the environment, permission is granted (see box, "Regulatory Oversight"). As of April 2005, APHIS had received 103 petitions for deregulation and had granted 63 (figure 4). Thirty-six percent of the released varieties have herbicide-tolerance traits, 27 percent have insect-resistance traits, and 17 percent have product-quality traits (figure 5).

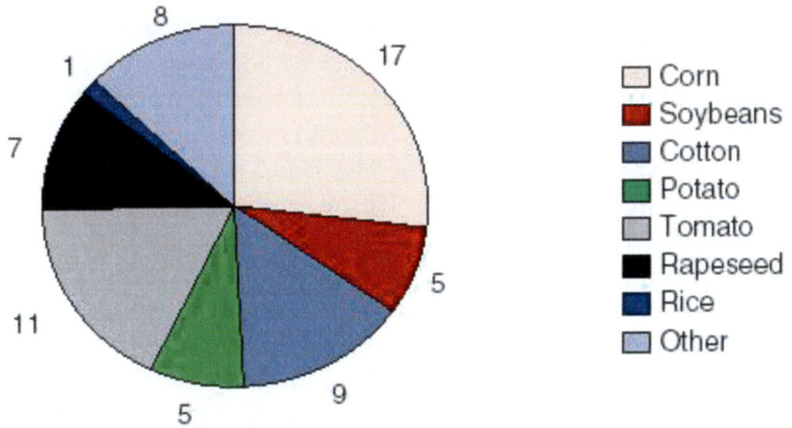

Source: Virginia Polytechnic Institute and State University, 2005.

Figure 4. Petitions for deregulation approved by APHIS, by crop

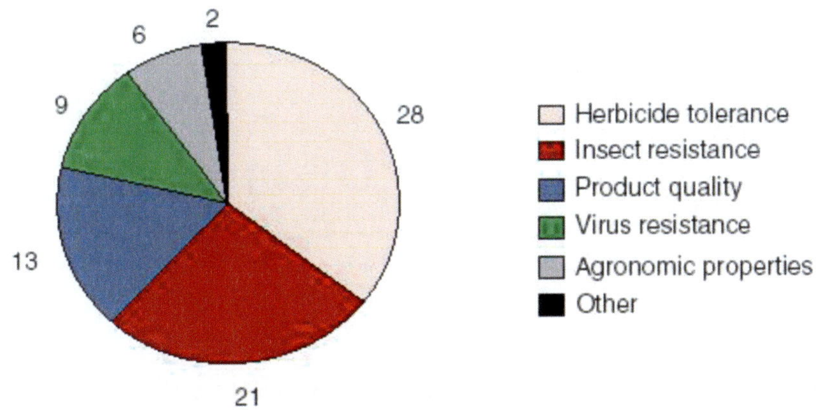

Source: Virginia Polytechnic Institute and State University, 2005.

Figure 5. Petitions for deregulation approved by APHIS, by GE trait

Adoption of GE Crops by U.S. Farmers Increases Steadily

Farmers are more likely to adopt new practices and technologies if they expect to benefit from them. Benefits are usually thought of in monetary terms, but can also include ease of operation, time savings, lower exposure to chemicals, and other factors. Farmers choose technologies and practices they expect to yield the greatest benefit based on their own preferences, farm characteristics, demand for their product, and costs.

Farmers' expectations of higher yields, savings in management time, and lower pesticide costs have driven a rapid increase in the adoption of GE crop varieties in the United States and several other countries. An estimated 200 million acres of GE crops with herbicide tolerance and/or insect resistance traits were cultivated in 17 countries worldwide in 2004, a 20-percent increase over 2003. U.S. acreage accounts for 59 percent of this amount, followed by Argentina (20 percent), Canada and Brazil (6 percent each), and China (5 percent) (ISAAA, 2004).[3]

GE varieties of soybeans, corn, and cotton have been available commercially in the United States since 1996, and the rate of adoption by U.S. farmers has climbed in most years since then (figure 6). For the most part, farmers have adopted herbicide-tolerant (HT) varieties—which help control weeds by enabling crops to survive certain herbicides that previously would have destroyed them along with the targeted weeds—at a faster pace than insect-resistant (Bt) varieties.

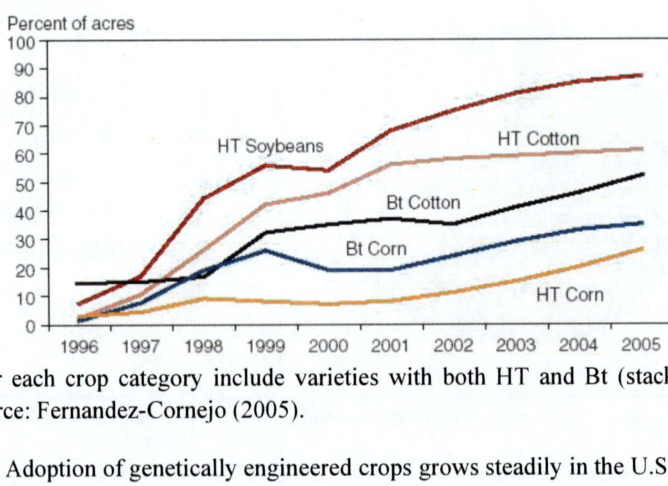

*Data for each crop category include varieties with both HT and Bt (stacked) traits.
Source: Fernandez-Cornejo (2005).

Figure 6. Adoption of genetically engineered crops grows steadily in the U.S.*

Weeds are such a pervasive pest for soybeans, corn, and cotton that over 90 percent of U.S. planted acreage for each crop has been treated with herbicides in recent years. The acreage share for HT soybeans has expanded more rapidly than that for HT varieties of cotton and corn, reaching 87 percent of U.S. soybean acreage in 2005.

Insect-resistant crops contain a gene from the soil bacterium *Bacillus thuringiensis* (Bt) that produces a protein toxic to specific insects. Acreage shares for Bt cotton and corn are lower than those for HT soybeans and cotton, and adoption is more concentrated in areas with a high level of infestation of targeted pests (insect infestation varies much more widely across locations than does weed infestation). Farmers planted Bt cotton to control tobacco budworm, bollworm, and pink bollworm on 52 percent of U.S. cotton acreage in 2005. Bt corn, originally developed to control the European corn borer, was planted on 35 percent of corn acreage in 2005, up from 24 percent in 2002. The recent increase in acreage share may be largely due to the commercial introduction in 2003/04 of a new Bt corn variety that is resistant to the corn rootworm, a pest that may be even more destructive to corn yield than the European corn borer (Comis).

Other GE crops planted by U.S. farmers over the past 10 years include HT canola, virus-resistant papaya, and virus-resistant squash (table 2). In addition, Bt potato varieties were introduced in 1996 but withdrawn from the market after the 2001 season, and a tomato variety genetically engineered to remain on the vine longer and ripen to full flavor after harvest was introduced in 1994 but was withdrawn from the market after being available sporadically for several years.

U.S. Farmers Expect to Profit from Adopting GE Crops

According to USDA's Agricultural and Resource Management Surveys (ARMS) conducted in 2001-03, most of the farmers adopting GE corn, cotton, and soybeans indicated that they did so mainly to increase yields through improved pest control (figure 7). Other popular reasons for adopting GE crops were to save management time and make other practices easier and to decrease pesticide costs. These results confirm other studies showing that expected profitability increases through higher yields and/or lower costs (operator labor, pesticides) positively influence the adoption of agricultural innovations.

Adoption of GE Crops and Yields

Currently available GE crops do not increase the yield potential of a hybrid variety. In fact, yield may even decrease if the varieties used to carry the herbicide-tolerant or insect-resistant genes are not the highest yielding cultivars.[4] However, by protecting the plant from certain pests, GE crops can prevent yield losses compared with non-GE hybrids, particularly when pest infestation is high. This effect is particularly important for Bt crops. For example, before the commercial introduction of Bt corn in 1996, the European corn borer was only partially controlled using chemical insecticides. Chemical use was not always profitable, and timely application was difficult. Many farmers accepted yield losses rather than incur the expense and uncertainty of chemical control. For those farmers, the use of Bt corn resulted in yield gains rather than pesticide savings. On the other hand, a recently introduced Bt corn trait selected for resistance against the corn rootworm, previously controlled using chemical insecticides, may provide substantial insecticide savings.[5]

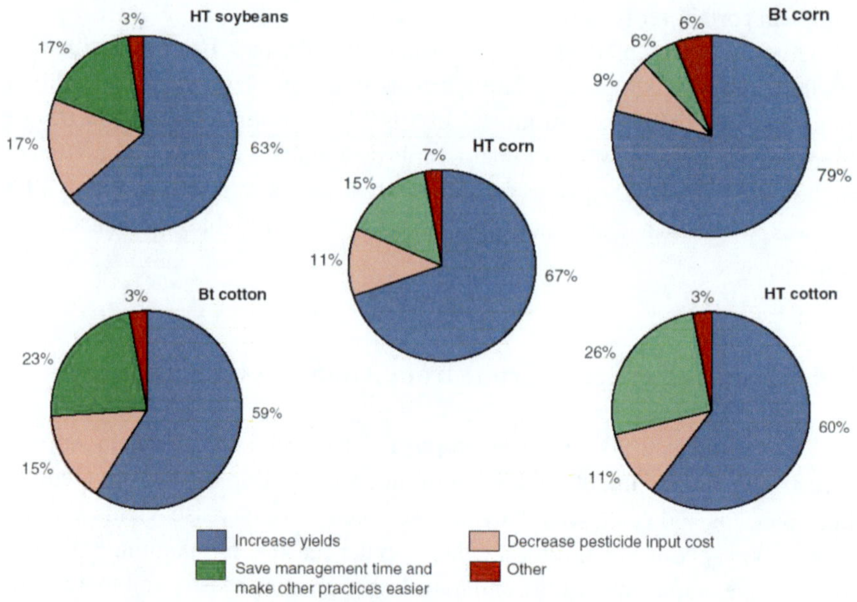

Source: Compiled by USDA's Economic Research Service using data from 2001, 2002, and 2003 Agricultural Resource Management Survey.

Figure 7. Farmers' reasons for adopting GE crops

Many field tests and farm surveys have examined the yield and cost effects of using GE crops (table 3). The majority of the results show GE crops produce higher yields than conventional crops.

A 2002 ERS study found that increases in cotton yields in the Southeast were associated with the adoption of HT and Bt cotton in 1997—a 10- percent increase in HT cotton acreage led to a 1.7-percent increase in yield and a 10-percent increase of Bt cotton acreage led to a 2.1-percent increase in yield. Increases in soybean yields associated with the adoption of HT soybeans were statistically significant but small (Fernandez-Cornejo and McBride, 2002).[6]

A more recent ERS study using 2001 survey data found that, on average, actual corn yield was 12.5 bushels per acre higher for Bt corn than for conventional corn, an increase of 9 percent (Fernandez-Cornejo and Li, 2005).[7]

Adoption and Net Returns, Household Income, and Pesticide Use

The impacts of GE crop adoption on U.S. farmers vary by crop and technology. Many studies have assessed the effects of the adoption of GE crops on returns and pesticide use, and the results of these studies are summarized in table 3. ERS researchers found that:

Planting HT cotton and HT corn was associated with increased producer net returns, but HT corn acreage was limited.[8] The limited acreage on which HT corn has been adopted is likely to be acreage with the greatest comparative advantage for this technology. The positive financial association with adoption may also be due to low premiums for HT corn seed relative to conventional varieties in an attempt to expand market share (Fernandez-Cornejo and McBride, 2002).

Adoption of Bt cotton and corn was associated with increased returns when pest pressures were high. The adoption of Bt cotton had a positive association with producer net returns in 1997, but the association was negative for Bt corn in 1998. This suggests that Bt corn may have been used on some acreage where the (ex post) value of protection against the European corn borer was lower than the premium paid for the Bt seed. Because pest infestations vary from one region to another and from one year to another, the economic benefits of Bt corn are likely to be greatest where pest pressures are most severe. Farmers must decide to use Bt corn before they know what the

European corn borer pest pressure will be that year, and damage caused by the European corn borer varies from year to year. Some farmers may have incorrectly forecast infestation levels, corn prices, and/or yield losses due to pest infestations, resulting in "overadoption." Also, producers may be willing to pay a premium for Bt corn because it reduces the risk of significant losses if higher-than-expected pest damage does occur (Fernandez-Cornejo and McBride, 2002).

Despite the rapid adoption of HT soybeans by U.S. farmers, no significant association with net farm returns was evident in 1997 or 1998. The lack of increased profitability for some farmers who adopted HT soybeans suggests that factors other than those included in traditional farm returns calculations may be driving adoption for these farmers. In particular, weed control may become simpler and require less management time, which allows growers of HT soybeans to control a wide range of weeds and makes harvest easier and faster. One important alternative use of management time is off-farm employment by farm operators and their spouses (FernandezCornejo and McBride, 2002).

Adoption of HT soybeans is associated with increased household income. Recent ERS research showed that adoption of HT soybeans was associated with a significant increase in off-farm household income for U.S. soybean farmers. On-farm household income is not significantly associated with adoption but total farm household income is significantly higher for adopters, suggesting that most managerial time saved by adopters is used in off-farm work (Fernandez-Cornejo et al., 2005).

Adoption of GE crops is associated with reduced pesticide use. Pesticide use rates (in terms of active ingredient) on corn and soybeans have declined since the introduction of GE corn and soybeans in 1996 (figure 8). In addition, ERS research suggests that, controlling for other factors, pesticide use declined with adoption. There was an overall reduction in pesticide use associated with the increased adoption of GE crops (Bt and HT cotton, HT corn, and HT soybeans combined, using 1997/1998 data), resulting in a significant reduction in potential exposure to pesticides (Fernandez-Cornejo and McBride, 2002). Overall pesticide use on corn, soybeans, and cotton declined by about 2.5 million pounds, despite the slight increase in the amount of herbicides applied to soybeans. In addition, glyphosate used on HT crops is less than one-third as toxic to humans, and not as likely to persist in the environment as the herbicides it replaces (Fernandez-Cornejo and McBride, 2002).

Table 3. Summary of primary studies on the effects of genetically engineered crops on yields, pesticide use, and returns

Crop/researchers/ date of publication	Data source	Effects on		
		Yield	Pesticide use	Returns
Herbicide-tolerant soybeans				
Delannay et al., 1995	Experiments	Same	na	na
Roberts et al., 1998	Experiments	Increase	Decrease	Increase
Arnold et al., 1998	Experiments	Increase	na	Increase
Marra et al., 1998	Survey	Increase	Decrease	Increase
Fernandez-Cornejo et al., 2002[1]	Survey	Small increase	Small increase	Same
McBride & El-Osta, 2002[2]	Survey	na	na	Same
Duffy, 2001	Survey	Small decrease	na	Same
Herbicide-tolerant cotton				
Vencill, 1996	Experiments	Same	na	na
Keeling et al., 1996	Experiments	Same	na	na
Goldman et al., 1998	Experiments	Same	na	na
Culpepper and York, 1998	Experiments	Same	Decrease	Same
Fernandez-Cornejo et al., 2000[1]	Survey	Increase	Same	Increase
Herbicide-tolerant corn				
Fernandez-Cornejo and Klotz-Ingram, 1998	Survey	Increase	Decrease	Same
McBride & El-Osta, 2002[2]	Survey	na	na	Increase
Bt cotton				
Stark, 1997	Survey	Increase	Decrease	Increase
Gibson et al., 1997	Survey	Increase	na	Increase
ReJesus et al., 1997	Experiments	Same	na	Increase
Bryant et al., 1999[3]	Experiments	Increase	na	Increase

Table 3. (Continued)

Crop/researchers/ date of publication	Data source	Effects on		
		Yield	Pesticide use	Returns
Marra et al., 1998	Survey	Increase	Decrease	Increase
Fernandez-Cornejo et al., 2000[1]	Survey	Increase	Decrease	Increase
Bt corn				
Rice and Pilcher, 1998	Survey	Increase	Decrease	Depends on infestation
Marra et al., 1998	Survey	Increase	Decrease	Increase
Benbrook, 2001	Survey	Increase	na	Decrease
McBride & El-Osta, 2002[2]	Survey	na	na	Decrease
Duffy, 2001	Survey	Increase	na	Same
Pilcher et al., 2002	Survey	Increase	Decrease	na
Baute, Sears, and Schaafsma, 2002	Experiments	Increase	na	Depends on infestation
Dillehay et al., 2004[4]	Experiments	Increase	na	na
Fernandez-Cornejo & Li, 2005[5]	Survey	Increase	Decrease	na

na = not analyzed in the study.

[1] Results using 1997 data.

[2] Results using 1998 data.

[3] Results are for 1996 and 1998. Results were different in 1997 when pest pressure was low.

[4] Results using 2000-2002 data.

[5] Results using 2001 data.

[6] Net returns equal revenues minus variable costs.

More recently, using 2001 data, ERS found that insecticide use was 8 percent lower per planted acre for adopters of Bt corn than for nonadopters (Fernandez-Cornejo and Li, 2005).[9]

The ERS results generally agree with field-test and other farm surveys that have examined the effects of using GE crops (table 3). The majority of those results show that pesticide use for adopters of GE crops is lower than for users of conventional varieties.

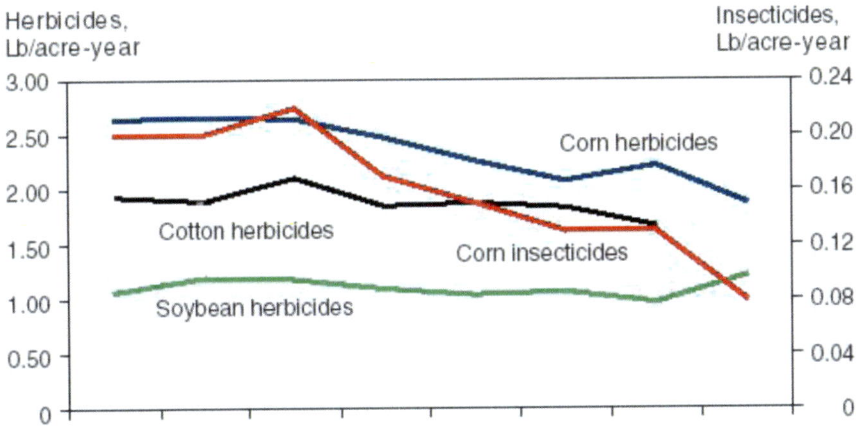

Source: NSS surveys.

Figure 8. Pesticide use in major field crops

Adoption of HT soybeans appears to be associated with conservation tillage. The environmental impact of conservation tillage is well documented.[10] The use of conservation tillage reduces soil erosion by wind and water, increases water retention, and reduces soil degradation and water and chemical runoff.

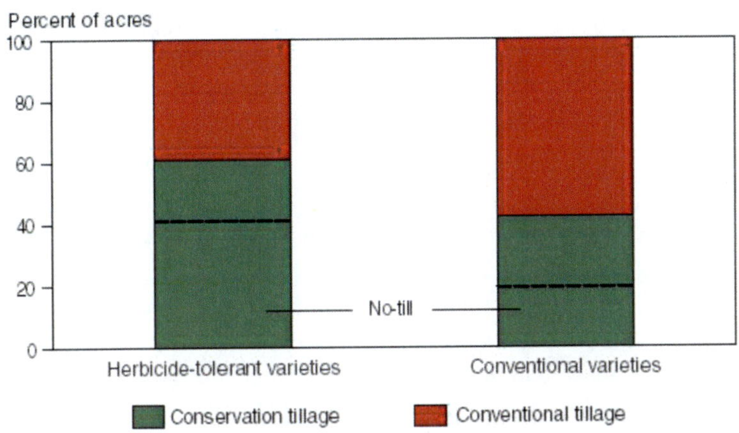

*Conservation tillage acres includes acres under no-till, ridge till, and mulch-till systems.

Source: Fernandez-Cornejo and McBride (2002).

Figure 9. Soybean area under conservation tillage* and no-till, 1997

According to USDA survey data, about 60 percent of the area planted with HT soybeans was under conservation tillage in 1997, compared with only about 40 percent of the acres planted with conventional soybeans (figure 9). Differences in the use of no-till between adopters and nonadopters of HT soybeans are even more pronounced: 40 percent of acres planted with HT soybeans were under no-till, twice the corresponding share of acreage planted with conventional soybeans. As a result, adoption of HT crops may indirectly benefit the environment by encouraging farmers to use soil conservation practices.

Consumer Demand Affects R&D, Adoption, and Marketing of GE-Derived Products

Investments in biotechnology-related research and development (R&D), the adoption of GE seeds, and the marketing of GE-derived products are all affected by consumer demand. While several surveys indicate that some U.S. consumers are concerned about GE food (table 4), these concerns have not had a large impact on the market for foods containing GE ingredients in the United States. In the European Union and a few other countries, consumer concern has resulted in substitution away from GE ingredients.

While opinion surveys give some indication of whether or not consumers are concerned about foods containing GE ingredients, they give little indication of the level of concern. Some researchers have attempted to quantify this concern through studies in which consumers are asked how much they would be willing to pay for foods made with GE ingredients, and for foods without GE ingredients. Researchers then use these data to measure whether or not there is a difference between these two hypothetical prices.

In most of these studies (table 5), consumers indicated that they were willing to pay more on average for GE-free foods or to avoid foods containing GE ingredients. However, in many of the studies, at least some consumers did not require a discount to buy foods containing GE ingredients, while some expressed that they would not be willing to buy foods containing GE ingredients at all.[11] Some respondents were willing to pay more for certain characteristics, such as improved nutrition and environmental benefits (Li et al., 2001; Lusk, 2003, Bocaletti and Moro, 2000).

Table 4. Surveys on Consumer Perceptions of Foods Containing GE Ingredients

Country/ Population	Surveyed by	Details
United States	Pew Initiative/ Mellman Group, 2003, 2004	27 percent favor introduction of GE foods;47 percent oppose. However, 64 percent disagree with the statement, "genetically modified foods should not be allowed to be sold even if the Food and Drug Administration believes they are safe,"and 28 percent feel that those foods should not be allowed, even if the FDA feels they are safe.
United States	Gallup, 2001	52 percent support the application of biotechnology; 38 percent oppose the use of biotechnology in food production.
United States	Hallman, 2004	47 percent approved or leaned toward approval of the use of GE to make plant-based foods, 41 percent disapproved or leaned toward disapproval, and 12 percent were unsure.
United States	IFIC, 2005	50 percent said likely to buy and 45 percent said not likely to buy GE produce modified to taste better or fresher;64 percent said likely to buy and 32 percent said not likely to buy GE produce modified to require fewer pesticide applications.
Beijing, China	Hu and Chen, 2004	67 percent were concerned about biotechnology.
Nanjing, China	Zhong et al., 2002	40 percent would buy GE foods;17 percent would not; 34 percent don't know.
Beijing, China, Shiajiazhuang, China	Ho and Vermeer, 2004	40 percent were willing or rather willing to consume foods containing GE-based ingredients, 51 percent were neutral, and 9 percent were rather unwilling or very unwilling to consume the foods.
Flemish speakers in Belgium	Verdurme and Viaene, 2003	15 percent opposed to GE foods;34 percent perceived small risks and small benefits;26 percent perceived moderate risks and moderate benefits;and 23 percent perceived large benefits.
United Kingdom	2003 GE Public Debate Steering board	86 percent preferred not to eat GE foods;8 percent happy to eat GE foods.

Source: Compiled by USDA's Economic Research Service.

Table 5. Willingness to Pay for Foods That do not Contain GE Ingredients[1]

Country	Food	Study	Willingness-to-pay premium
United States	Vegetable oil	Tegene et al., 2003	In experimental auctions, consumers willing to pay 14 percent more for non-GE food.
United States	Potatoes	Loureiro and Hine, ` 2002	Customers willing to pay 5 percent more for non-GE food.
United States	Golden rice	Lusk, 2003	Customers willing to pay 93 cents for GE "golden rice"with added vitamin C, 65-75 cents for regular rice.
United Kingdom	All foods	Burton et al., 2001	Customers indicated willingness to increase food budgets by 26-129 percent to avoid GE foods.
Italy	*	Bocaletti and Moro, 2000	Consumers willing to pay a positive amount for GE attributes;66 percent did not require a premium to consume GE foods.
United States, France, Germany, and United Kingdom	Beef fed with GE feed	Lusk et al., 2003	U.S.consumers willing to pay $2.83 and $3.31 per lb.to avoid GE;European consumers $4.86 to $11.01.
United States, United Kingdom	Breakfast cereal	Moon and Balasubramanian, 2001	Survey found 56 percent of UK consumers willing to pay a premium to avoid GE compared with 37 percent of U.S.consumers.
Norway, United States, Japan, Taiwan	Vegetable oil	Chern et al., 2002	Norwegian students were willing to pay $1.51 (55-69 percent premium) per liter for non-GE vegetable oil, U.S.students were willing to pay $1.13 (50-62 percent premium), Japanese students were willing to pay $0.88 (33-40 percent premium), and Taiwanese students were willing to pay $0.45 cents (17-21 percent premium).

Table 5. (Continued)

Country	Food	Study	Willingness-to-pay premium
China	Rice	Li et al., 2002	80 percent of consumers did not require a premium to purchase GE rice and on average were willing to pay a 38-percent premium on GE rice and a 16-percent premium for GE soy oil.
Norway	Bread	Grimsrud, et al., 2004	Consumers required discounts of 37-63 percent to buy GE bread;One-fourth willing to buy with no discount.
Australia	Beer	Burton and Pearse, 2002	Younger consumers would pay $A 0.72 less and older consumers $A 0.40 less for beer made with GE barley.
Canada	*	West et al., 2002	83 percent of consumers ascribed a lower value to several GE foods.
France	*	Noussair et al., 2004	35 percent of consumers were unwilling to purchase GE foods, and 42 percent were willing to purchase them if they were less expensive.
United States	Oil, chips, and potatoes	Rousu et al., 2004	Consumers reduced their demand by an average of 7-13 percent for each food product having 1 percent and 5 percent tolerance levels for GE material relative to GE-free food.

[1]See also Lusk et al. (2005), who summarize a set of 25 studies including 57 GE valuation studies and report that, on average, consumers are willing to pay a positive premium for GE-free foods.

*This study did not focus on a specific food item.

Source: Compiled by USDA's Economic Research Service.

While surveys and willingness-to-pay studies provide some insight into consumer opinion, they often do not reflect how consumers will behave in a real market situation when purchasing goods and services. Each food product has many characteristics, such as taste, color, and ripeness. The presence of a biotech-derived component is only one attribute. Empirically, it is difficult to determine what percentage of the price a consumer is paying for a specific characteristic. There are no published studies that indicate how many consumers have actually paid a premium to purchase non-GE goods, but there is some empirical evidence of the types of goods that are currently offered for sale to consumers. In the United States, many products contain GE ingredients, and the demands for these products apparently have been unaffected by negative opinions about biotechnology expressed in surveys. A few specialty brands are marketed as "GE free," but they represent a small percentage of supermarket sales.[12] In some other countries, however, strong consumer demand for non-GE products has limited the availability of GE items (see box, "Biotech Product Differentiation: A Tale of Two Markets").

ADOPTION OFFERS MARKET BENEFITS TO MANY STAKEHOLDERS

In addition to farmers, seed suppliers, technology providers, and consumers also benefit from the adoption of GE crops in the United States. Biotechnology developers and seed firms benefit by charging technology fees and seed premiums to adopters of GE varieties. U.S. and foreign consumers may benefit indirectly from GE crops through lower commodity prices that result from increased supplies.[13]

ERS estimated the total market benefit arising from the adoption of three GE crops in the United States—HT soybeans, Bt cotton, and HT cotton—in 1997 (Price et al., 2003).[14] Estimated benefits to farmers, seed producers, and consumers were around $210 million for Bt cotton, $230 million for HT cotton, and $310 million for HT soybeans. This estimate includes the change in total welfare in both the seed input and commodity output markets. The distribution of these benefits among consumers, farmers, technology providers (biotech firms), seed firms, and consumers and producers in the rest of the world (ROW) is shown in figures 10-12. The distribution of benefits varies by crop and technology because the economic incentives to farmers (crop prices and production costs), the payments to technology providers (biotech firms) and seed firms, and the effect of the technology on world crop prices are different for each crop and technology. For example, adoption of HT cotton benefits mainly consumers while Bt cotton benefits farmers and technology providers. Seed firms are by far the largest beneficiaries in the case of soybeans.

These results should be interpreted carefully, since the estimates are based on only a few years of data. Moreover, estimated benefits and their distribution depend particularly on the analytical framework, supply and demand elasticity assumptions,[15] crops considered, and year-specific factors (such as weather).

In particular, the benefits attributable to HT soybeans and their distribution are very dependent on the soybean supply elasticity. Table 6 shows estimates of the benefits of Bt cotton and HT soybeans and their distribution obtained by other researchers.

BIOTECH PRODUCT DIFFERENTIATION: A TALE OF TWO MARKETS

The introduction of genetically engineered (GE) crops has led food manufacturers to make a choice for each of their products: either pursue a non-GE strategy and market and produce a non-GE product or source inputs based on cost and quality, or market and produce an undifferentiated product.

If all manufacturers were to pursue a non-GE strategy, farmers would eventually abandon GE technologies and consumer choice would be restricted to potentially higher cost non-GE products. If manufacturers were to pursue an undifferentiated strategy, then farmers' use of the technology would be determined by production costs and consumers would be faced with markets in which they could not differentiate between GE and non-GE foods. If manufacturers pursue both strategies, some farmers would continue to use the technology while others would grow conventional crops to supply non-GE markets. In this scenario, consumers would have a choice between GE and non-GE food, at least for some products.

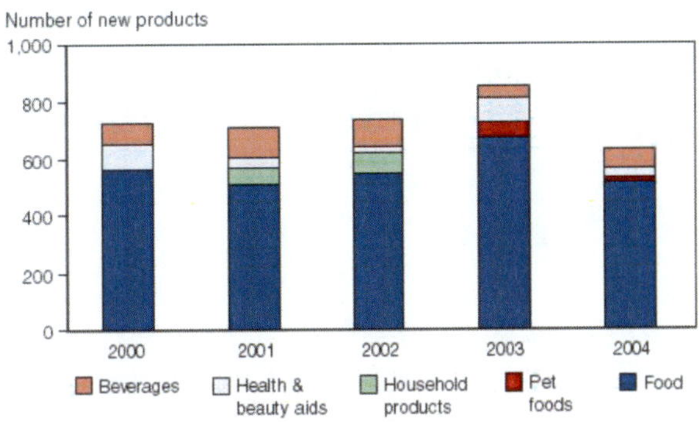

Source: Productscan Online.

Annual non-GE new product introductions in the United States

In the United States, where unlabeled foods may contain GE ingredients, the data show that manufacturers have been active in creating a market for GE- free foods. From 2000 to 2004, manufacturers introduced over 3,500 products that had explicit non-GE labeling, mostly food products, with annual totals ranging from 854 in 2003 to 631 in 2004. This is in addition to organic foods (organic crops may not be grown using GE techniques) (Dimitri and Greene, 2002).

In the European Union and Japan, where unlabeled foods cannot contain GE ingredients, manufacturers have chosen a non-GE marketing strategy. Very few products labeled as containing GE ingredients are found on European or Japanese grocery store shelves.

The data also show that there have been limited attempts to market GE products in the United States. There were far fewer new GE products introduced than new non-GE products, and most of the GE products were introduced in the 1 990s. GE products included tomatoes (advertised as better tasting with a longer shelf life), canola oil (advertised as heat stable), shrimp (advertised as gourmet-quality), beef (low-fat), dietary supplements, cigarettes (low-nicotine), and a drain cleaner.

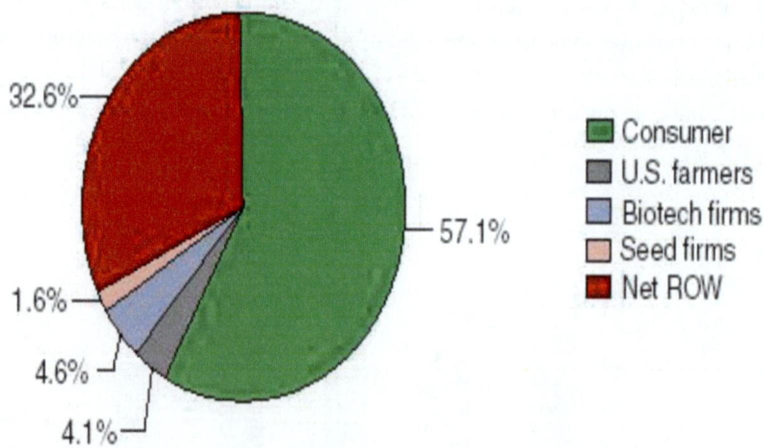

Source: Price et al., 2003.

Figure 10. Stakeholders' shares of the estimated total world benefit from adopting herbicide-tolerant cotton, 1997

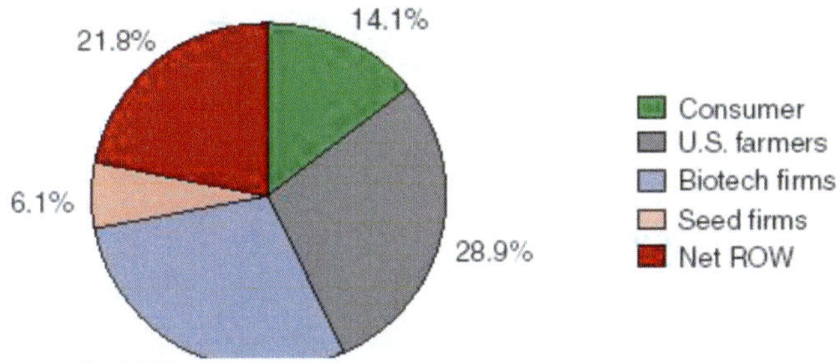

Source: Price et al., 2003.

Figure 11. Stakeholders' shares of the estimated total world benefit from adopting Bt-cotton, 1997

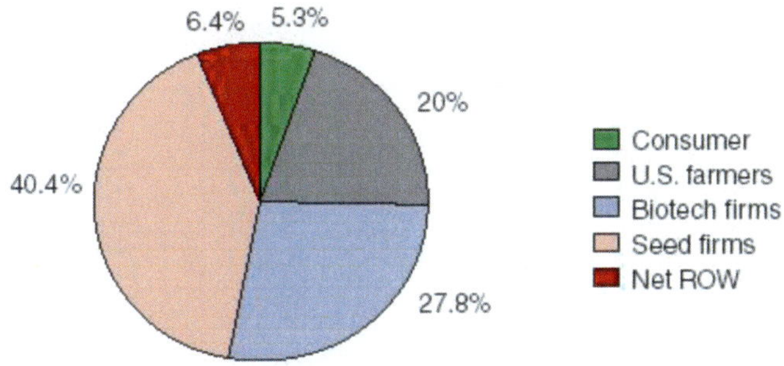

Source: Price et al., 2003.

Figure 12. Stakeholders' shares of the estimated total world benefit from adopting herbicide-tolerant soybeans, 1997

CONCLUSION

The role that biotechnology plays in agriculture in the United States and globally depends on a number of factors and uncertainties. As the USDA Advisory Committee on Biotechnology and 21st Century Agriculture report indicates, "agricultural biotechnology sits at the crossroads of other debates on the future of American and world agriculture, on international trade relations, on biological diversity and the development of international instruments related to its preservation and exploitation, on the role of the multinational corporations, and on how best to build public confidence in rapidly evolving emerging technologies in general" (p.2.). One thing seems certain, however:

the ultimate contribution of agricultural biotechnology will depend on our ability to identify and measure its potential benefits and its risks as well as their distribution.

Table 6. Benefits of GE Techniques and Their Distribution (from Estimates in Related Studies)

Study	Year	Total benefits	Share of the total benefits			
			U.S. farmers	Innovators	U.S. consumers	Net ROW
		$ million	*Percent*			
Bt cotton						
Falck-Zepeda et al. (1999)	1996	134	43	47	6	
Falck-Zepeda et al. (2000b)	1996	240	59	26	9	6
Falck-Zepeda et al. (2000a)	1997	190	43	44	7	6
Falck-Zepeda et al. (1999)	1998	213	46	43	7	4
Frisvold et al. (2000)	1996-98	131-164	5-6	46	33	18
EPA (2000)[1]	1996-99	16.2-45.9	n.a.	n.a.	n.a.	n.a.
Price et al. (2003)	1997	210	29	35	14	22
Herbicide-tolerant soybeans						
Falck-Zepeda et al. (2000a)	1997-Low elasticity[2]	1,100	77	10	4	9
	1997-High elasticity[3]	437	29	18	17	28
Moschini et al. (2000)	1999	804	20	45	10	26
Price et al. (2003)	1997	310	20	68	5	6

n.a. = Not applicable.

ROW = Rest of the world.

[1]Limited to U.S. farmers.

[2]Assumes a U.S. soybean supply elasticity of 0.22.

[3]Assumes a U.S. soybean supply elasticity of 0.92.

Source: Price et al., 2003.

REFERENCES

Arnold, J. C., Shaw, D. R. & Medlin, C.R. (1998). "Roundup Ready™ Programs Versus Conventional Programs: Efficacy, Varietal Performance, and Economics." *Proceedings of the Southern Weed Science Society, Southern Weed Science Society, 51*, 272-73.

Associated Press & Matt Moore. (2004). Covered by CBS news as "Europeans Balk At Biotech Beer" July 8, 2004. http://www.cbsnews.com/stories/2004/07/08/tech/main628305.shtml.

Baker, G. & Burnham, T. (2001). "Consumer Response to Genetically Modified Foods: Market Segment Analysis and Implications for Producers and Policy Makers," *Journal of Agricultural and Resource Economics, 26*(2), (December), 387-403.

Baute, T.S., Sears, M. K. & Schaafsma, A.W. (2002). "Use of Transgenic *Bacillus thuringiensis* Berliner Corn Hybrids to Determine the Direct Economic Impact of the European Corn Borer (*Lepidoptera: Crambidae*) on Field Corn in Eastern Canada. *Journal of Economic Entomology. 95*(1), 57-64.

Boccaletti, S. & Moro, D. (2000). "Consumer Willingness-To-Pay For GM Food Products In Italy," *AgBioForum*, *3*(4), 259-267.

Bryant, K.J., W.C. Robertson, and G.M. Lorenz III. (1998). "Economic Evaluation of Bollgard Cotton in Arkansas." In *Proceedings of the Beltwide Cotton Conferences, National Cotton Council, Memphis, TN. 1,* 388-89.

Burton, M. & Pearse, D. (2002). "Consumer Attitudes Towards Genetic Modification, Functional Foods, and Microorganisms: A Choice Modeling Experiment for Beer." *AgBioForum*, *5*(2), 51-58.

Burton, M., Rigby, D., Young, T. & James, S. (2001). "Consumer Attitudes to Genetically Modified Organisms in Food in the UK." *European Review of Agricultural Economics*, *28* (4),(December), 479-98.

Carter, C.A., & Gruère, G. P. (2003(a)). "Mandatory Labeling of Genetically Modified Foods: Does it Really Provide Consumer Choice?" *AgBioForum*, *6*(1 & 2), 68-70.

Carter, C. A., & Gruère, G. P. (2003(b)). "International Approaches to the Labeling of Genetically Modified Foods." *Choices*, June.

Chern, W. S., Rickertsen, K., Tsuboi, N. & Tsu-Tan Fu. (2002). "Consumer Acceptance and Willingness to Pay for Genetically Modified Vegetable Oil and Salmon: A Multiple-Country Assessment." *AgBioForum*, *5*(3), 105-112.

Colorado State University. (2004). Department of Soil and Crop Sciences. "Discontinued Transgenic Products," in *Transgenic Crops: An Introduction and Resource Guide*. http://www.colostate.edu/programs/lifesciences/TransgenicCrops/defunct.html. March.

Comis, D. (1997). "*Safe Corn Pest Bait Expected to Slash U.S. Insecticide Use*." U.S. Department of Agriculture, Agricultural Research Service. http://www.ars.usda.gov/is/pr/1997/970116.htm?pf=1.

Conservation Technology Information Center (CTIC). National Crop Residue Management Survey. "Conservation Tillage Data." http://www.ctic.purdue.edu/CTIC/CRM.html.

Culpepper, A.S. & York, A. C. (1998). "Weed Management in Glyphosate-Tolerant Cotton." *The Journal of Cotton Science, 4*,174-185.

Delannay, X., Bauman, T. T., Beighley, D. H., Buettner, M. J., Coble, H. D., DeFelice, M. S., Derting, C. W., Diedrick, T. J., Griffin, J. L., Hagood, E. S., Hancock, F. G., Hart, S. E., LaVallee, B. J., Loux, M. M., Lueschen, W. E., Matson, K. W., Moots, C. K., Murdock, E., Nickell, A. D., Owen, M. D. K., Paschall II, E. H., Prochaska, L. M., Raymond, P. J., Reynolds, D. B., Rhodes, W. K., Roeth, F. W., Sprankle, P. L., Tarochionc, L. J., Tinius, C. N., Walker, R.H., Wax, L. M., Weigelt, H. D. & Padgette. S. R. (1995). "Yield Evaluation of a GlyphosateTolerant Soybean Line After Treatment with Glyphosate." *Crop Science*, *35*, 1461-67.

Dillehay, B. L., Roth, G. W., Calvin, D. D., Karatochvil, R. J., Kuldau, G. A. & Hyde, J. A. "Performance of Bt Corn Hybrids, their Near Isolines, and Leading Corn Hybrids in Pennsylvania and Maryland." *Agronomy Journal*, *96*(204), 818-824.

Dimitri, C. & Greene, C. (2002). *Recent Growth Patterns in the U.S. Organic Foods Market.* U.S. Department of Agriculture, Economic Research Service. Agriculture Information Bulletin No. 777. Sept.

Duffy, M. (2001). "*Who Benefits from Biotehnology*?" Paper presented at the American Seed Trade Association meeting. Chicago, IL. December 5-7.

Falck-Zepeda, J. B., Traxler, G. & Nelson, R. G. (1999). "*Rent Creation and Distribution from the First Three Years of Planting Bt Cotton*." ISAAA Briefs No. 14. International Service for the Acquisition of Agri-biotech Applications. Ithaca, NY.

Falck-Zepeda, J. B., Traxler, G. & Nelson, R. G. (2000a). "Rent Creation and Distribution from Biotechnology Innovations: The Case of Bt Cotton and Herbicide-Tolerant Soybeans in 1997," *Agribusiness, 16*, 21-32.

Falck-Zepeda, J. B., Traxler, G. & Nelson, R.G. (2000b). "Surplus Distribution from the Introduction of a Biotechnology Introduction." *American Journal of Agricultural Economics*, *82,* 360-369.

Fernandez-Cornejo, J. (2004). *The Seed Industry in U.S. Agriculture: An Exploration of Data and Information on Crop Seed Markets, Regulation, Industry Structure, and Research and Development.* U.S. Department of Agriculture, Economic Research Service, Agriculture Information Bulletin No. 786. Feb.

Fernandez-Cornejo, J. (2005). "*Adoption of Genetically Engineered Crops in the U.S.*" Data product. U.S. Department of Agriculture, Economic Research Service. http://www.ers.usda.gov/Data/BiotechCrops/. July.

Fernandez-Cornejo, J. & Jiayi Li. (2005). "The Impacts of Adopting Genetically Engineered Crops in the USA: The Case of Bt Corn." Paper presented at the American Agricultural Economics Association meetings. Providence, RI..

Fernandez-Cornejo, J. & McBride, W. D. (2000). *Genetically Engineered Crops for Pest Management in U.S. Agriculture*. U.S. Department of Agriculture, Economic Research Service, Agricultural Economic Report No. 786. April.

Fernandez-Cornejo, J. & McBride, W. D. (2002). *The Adoption of Bioengineered Crops*. U.S. Department of Agriculture, Economic Research Service, Agricultural Economic Report No. 810. May.

Fernandez-Cornejo, J., Hendricks, C. & Mishra, A. (2005). "Technology Adoption and Off-Farm Household Income: The Case of Herbicide-Tolerant Soybeans." *Journal of Agricultural and Applied Economics*, *37*(2).

Fernandez-Cornejo, J., Klotz-Ingram, C. & Jans, S. (2000). "Farm-Level Effects of Adopting Genetically Engineered Crops in the U.S.A.," In *Transitions in Agrobiotech: Economics of Strategy and Policy*, pp. 57-72. W.L. Lesser, editor. Food Marketing Research Center, University of Connecticut and Department of Research Economics, University of Massachusetts.

Fernandez-Cornejo, J., Klotz-Ingram, C. & Jans, S. (2002). "Farm-Level Effects of Adopting Herbicide-Tolerant Soybeans in the U.S.A." *Journal of Agricultural and Applied Economics*, *34*(1), 149-163.

Fernandez-Cornejo, J., Caswell, M. & Klotz-Ingram, C. (1999). "Seeds of Change: From Hybrids to Genetically Modified Foods." *Choices,* (Millennium issue, 4th quarter of): 18-22.

Finke, M. S. & Kim, H. (2003). "Attitudes About Genetically Modified Foods Among Korean and American College Students." *AgBioForum*, *6*(4), 19 1-197.

Frisvold, G., Tronstad, R. & Mortensen, J. (2000). "*Effects of Bt Cotton Adoption: Regional Differences and Commodity Program Effects*." Paper presented at the Western Agricultural Economics Association Meeting, Vancouver, British Columbia. June 30.

Gibson, J. W. IV, Laughlin, D., Lutrell, R. G., Parker, D., Reed, J. & Harris, A. (1997). "Comparison of Costs and Returns Associated with *Heliothis* Resistant Bt Cotton to Non-Resistant Varieties," *Proceedings of the Beltwide Cotton Conferences. Cotton Weed Science Research Conference. v.1,* pp. 244-247.

Goldman, D.L., Guy, C., McCelland, M. & Kendig, A. (1996). "Weed Control Systems in Roundup Ready™ Cotton." In *Proceedings of the Beltwide Cotton Conferences. Cotton Weed Science Research Conference. v.2*. p. 1532.

Gracia, A. & Albisu, L. M. (2001). "Food Consumption in the European Union: Main Determinants and Country Differences." *Agribusiness, 17*(4),(Autumn), 469-488.

Graeber, J. V., Nafziger, E. D. & Mies, D.W. (1999). "Evaluation of Transgenic, Bt- Containing Corn Hybrids." *Journal of Production Agriculture, (12)(4)*, 659-663.

Greenpeace. EU Markets. (2005). http://www.greenpeace.org/raw/content/international/press/reports/european-markets. February 3.

Grimsrud, K. M., McCluskey, J., Loureiro, M. & Wahl, T. (2004). "Consumer Attitudes to Genetically Modified Foods in Norway." *Journal of Agricultural Economics*, *55*(1),(March), 75-90.

Hallman, W., Hebden, W., Cuite, C., Aquino, H. & Lang, J. (2004). *Americans and GM Food: Knowledge, Opinion & Interest in 2004*. Food Policy Institute. http://www.foodpolicyinstitute.org/docs/reports/NationalStudy.pdf.

Harrison, R. W., Boccaletti, S. & House, L. (2004). "Risk Perceptions of Urban Italian and United States Consumers for Genetically Modified Foods." *AgBioForum*, 7(4), 195-201.

Ho, P. & Vermeer, E. B. (2004). "Food Safety Concerns and Biotechnology: Consumers' Attitudes to Genetically Modified Products in Urban China." *AgBioFo rum*, 7(4), 158-175.

House, L., Lusk, J., Jaeger, S., Traill, W. B., Moore, M., Valli, C., Morrow, B. & Yee, W. M. S. (2004). "Objective and Subjective Knowledge: Impacts on Consumer Demand for Genetically Modified Foods in the United States and the European Union." *AgBioForum*, *7*(3), 113-123.

Hu, W. Y. & Chen, K. (2004). "Can Chinese Consumers Be Persuaded? The Case of Genetically Modified Vegetable Oil." *AgBioForum*, 7(3), 24-132.

International Food Information Council (IFIC) Polls, 1999-2005. http://ific.org/publications/.

International Service for the Acquisition of Agri-biotech Applications (ISAAA). Preview: Global Status of Commercialized Biotech/GM Crops: ISAAA Briefs 32-2004. http://www.isaaa.org/kc/CBTNews/press_release/briefs32/ESummary/Executive%20Summary%20(English).pdf.

Just, R. E. & Hueth, D. L. (1993). "Multimarket Exploitation: The Case of Biotechnology and Chemicals." *American Journal of Agricultural Economics, 75*(4), 936-45.

Kaufman, P. (2000). "Consolidation in Food Retailing: Prospects for Consumers and Grocery Suppliers." *Agricultural Outlook*, August.

Keeling, W., Dotray, P., Jones, C. & Sunderland, S. (1996). "Roundup™ Ready Cotton: A Potential New Weed Management Tool for the Texas High Plains." *Proceedings of the Beltwide Cotton Conferences. Cotton Weed Science Research Conference. v.2*. p. 1529.

King, J. & Heisey, P. (2003). "Ag Biotech Patents: Who is Doing What?" *Amber Waves*, November. http://www.ers.usda.gov/Amberwaves/November03/DataFeature/.

Li, Q., Curtis, K. R., McCluskey, J. J. & Wahl, T.I. (2002). "Consumer Attitudes Toward Genetically Modified Foods in Beijing, China." *AgBioForum*, *5*(4), 145-152.

List, J. A. & Gallet, C. (2001). "What Experimental Protocol Influence Disparities between Actual and Hypothetical Stated Values." *Environmental and Resource Economics, 20,* 241-254.

Loureiro, M. L. & Hine, S. (2002). "Discovering Niche Markets: A Comparison of Consumer Willingness to Pay for Local (Colorado Grown), Organic, and GMO-Free Products." *Journal of Agricultural and Applied Economics*, *34*(3), (December), 477-487.

Lusk, J. L. (2003). "Effects of Cheap Talk on Willingness to Pay for Golden Rice." *American Journal of Agricultural Economics*, *85*(4), (November), 840-856.

Lusk, J. L., Roosen, J. & Fox, J. (2003). "Demand for Beef with Growth Hormones and Fed Genetic Corn." *American Journal of Agricultural Economics*, *85* (1), (February), 16-29.

Lusk, J. L., Jamal, M., Kurlander, L., Roucan, M. & Taulman, L. (2005). "A Meta Analysis of Genetically Modified Food Valuation Studies." *Journal of Agricultural and Resource Economics*, *30*(April), 28-44.

Marks, L. A., Kalaitzandonakes, N. & Vickner, S. S. (2004). "Consumer purchasing behavior toward GM foods in the Netherlands," in *Consumer Acceptance of Biotech Foods*. R. Evenson and V. Santaniello (editors). Wallingford, UK: CABI Publishers, p. 23-29.

Marra, M., Carlson, G. & Hubbell, B. (1998). "Economic Impacts of the First Crop Biotechnologies." Available at http://www.ag.econ.ncsu.edu/faculty/marra/firstcrop/imp001.gif.

McBride, W. D. & El-Osta, H. (2002). "Impacts of the Adoption of Genetically Engineered Crops on Farm Financial Performance." *Journal of Agricultural and Applied Economics*, *Vol. 34*, No. 1(April), 175-19 1.

McBride, W., Short, S. & El-Osta, H. (2002). "*Production and Financial Impacts of the Adoption of Bovine Somatropin on U.S. Dairy Farms*." Paper presented at the 2002 American Agricultural Economics Association meetings, Long Beach, CA, July 28-3 1.

Moon, W. K. & Balasubramanian, S. K. (2001). "Public Perceptions And Willingness-To-Pay A Premium For Non-GM Foods in the US and UK." *AgBioForum*, *4*(3&4), 221-231.

Moon, W. K. & Balasubramanian, S. K. (2003). "Is There a Market for Genetically Modified Foods in Europe? Contingent Valuation of GM and Non-GM Breakfast Cereals in the United Kingdom." *AgBioForum*, *6*(3), 128-133.

Moschini, G., Lapan, H. & Sobolevsky, A. (2000). "Roundup Ready™ Soybeans and Welfare Effects in the Soybean Complex." *Agribusiness*, *16*, 3 3-55.

National Research Council. (2002). *Animal Biotechnology: Science Based Concerns*. Board on Agriculture and Natural Resources, Board on Life Sciences, Washington, DC: The National Academies Press. http://www.nap.edu/openbook/0309084393/html/R1.html#pagetop. (a).

National Research Council. (2002(b)). *Environmental Effects of Transgenic Plants: The Scope and Adequacy of Regulation*. Board on Agriculture and Natural Resources, Committee on Environmental Impacts Associated with Commercialization of Transgenic Plants. Washington, DC: The National Academies Press. http://www.nap.edu/books/0309082633/html/.

Noussair, C., Robin, S. & Ruffieux, B. (2004). "Do Consumers Really Refuse To Buy Genetically Modified Food?" *The Economic Journal*, *114*(492), 102-120.

North Carolina Farm Bureau Federation. (2001). "Gallup Conducts Poll on Biotech Food." Aug. 6-12.

Office of Science & Technology Policy. (1986). Executive Office of the President. *Coordinated Framework for Regulation of Biotechnology.* Federal Register, 51 FR 23302, http://usbiotechreg.nbii.gov/Coordinated_Framework_1986_Federal_Register.html.

Onyango, B., Nayga, Jr. R. M. & Schilling, B. (2004). "Role of Product Benefits and Potential Risks in Consumer Acceptance of Genetically Modified Foods." *AgBioForum*, *7*(4), 202-211.

Panos. (1998). "Greed or Need? Genetically Modified Crops." Panos Media Briefing No. 30. October. http://www.oneworld.org/panos/.

Pew Initiative on Food & Biotechnology. (2001). *Harvest on the Horizon: Future Uses of Agricultural Biotechnology.* http://pewagbiotech.org/research/harvest/harvest.pdf. Sept..

Pew Initiative on Food & Biotechnology. (2003). "*Public Sentiment About Genetically Modified Food: September Update.*" http://pewagbiotech.org/research/2003update/. 2003.

Pew Initiative on Food and Biotechnology. (2004). "Overview of Findings Focus Groups & Poll." http://pewagbiotech.org/research/2004update/overview.pdf. 2004.

Pilcher, C.D., Rice, M. E., Higgins, R. A., Steffey, K. L., Hellmich, R. L., Witkowski, J., Calvin, D., Ostlie, K. R. & Gray, M. (2002). "Biotechnology and the European Corn Borer: Measuring Historical Farmer Perceptions and Adoption of Transgenic Bt Corn as a Pest Management Strategy." *Journal of Economic Entomology, 95*(5), 878-892.

Price, G. K., Lin, W., Falck-Zepeda, J. B. & Fernandez-Cornejo, J. (2003). *The Size and Distribution of Market Benefits from Adopting Agricultural Biotechnology.* U.S. Department of Agriculture, Economic Research Service, Technical Bulletin No. 1906. Nov.

ReJesus, R. M., Greene, J. K. Hamming, M. D. & Curtis, C. E. (1997). "Economic Analysis of Insect Management Strategies for Transgenic Bt Cotton Production in South Carolina." *Proceedings of the Beltwide Cotton Conferences. Cotton Economics and Marketing Conference. v.1.* p. 247-251.

Rice, M. E. & Pilcher, C. D. (1998). "Potential Benefits and Limitations of Trans- genic Bt Corn for Management of the European Corn Borer (*Lepidoptera: Crambidae*)." *American Entomologist, 44,* 75-78.

Roberts, R. K., Pendergrass, R. & Hayes, R. M. (1998). "Farm-Level Economic Analysis of Roundup Ready™ Soybeans." Paper presented at the Southern Agricultural Economics Association Meeting, Little Rock, AR, Feb. 1-4.

Rousu, M., Huffman, W. E., Shogren, J. F. & Tegene, A. (2004). "Are United States Consumers Tolerant of Genetically Modified Foods?" *Review of Agricultural Economics*, 26(February), 19-31.

Runge, C. Ford & Ryan, B. (2004). *The Global Diffusion of Plant Biotechnology: International Adoption and Research in 2004*. University of Minnesota.
http://www.apec.umn.edu/faculty/frunge/globalbiotech04.pdf. Dec. 8.

Shoemaker, R., Harwood, J., Day-Rubenstein, K., Dunahay, T., Heisey, P., Hoffman, L., Klotz-Ingram, C., Lin, W., Mitchell, L., McBride, W. & Fernandez-Cornejo, J. (2001). *Economic Issues in Agricultural Biotechnology*. U.S. Department of Agriculture, Economic Research Service, Agriculture Information Bulletin No. 762. Feb.

Shoemaker, R., Johnson, D. & Golan, E. (2003). "Consumers and the Future of Biotech Foods in the United States." *Amber Waves*, November. http://www.ers.usda.gov/Amberwaves/November03/Features/futureof-biotech.htm.

Steering Board of the Public Debate on GM. *GM Nation?* (2003). *The Findings of the Public Debate*.
http://www.gmnation.org.uk/docs/gmnation_finalreport.pdf.

Tegene, A., W. Huffman, M. Rousu, and J. Shogren. (2003). *The Effects of Information on Consumer Demand for Biotech Foods: Evidence from Experimental Auctions*. U.S. Department of Agriculture, Economic Research Service, Technical Bulletin No. 1903. April.

Traill, W. B., Jaeger, S. R., Yee, W. M. S., Valli, C., House, L. O., Lusk, J. L. , Moore, M. & Morrow, Jr. J. L. (2004). "Categories of GM Risk-Benefit Perceptions and Their Antecedents." *AgBioForum*, *7*(4), 176-186.

Tripp, R. (1998). "Regulatory Issues: Varietal Registration and Seed Quality Control." In Morris, M.L (editor), *Maize Seed Industries in Developing Countries*. Boulder, CO: CIMMYT (International Maize and Wheat Improvement Center). pp. 159-173.

United Nations Conference on Trade and Development (UNCTAD). (2005). *International Trade in GMOs and GM Products: A National and*

Multilateral Legal Framework. Policy Issues in International Trade and Commodities Study Series No. 29. United Nations, New York, NY, and Geneva, Switzerland.

University of California. (1998). Davis Vegetable Research and Information Center, Pest-Resistance Management. "Insecticidal Bt-Potatoes." http://ucce.ucdavis.edu/files/filelibrary/5283/2196.pdf. Feb.

USDA Advisory Committee on Biotechnology and 21st Century Agriculture (AC21). (2005). *AC21 Consensus Report—Preparing for the Future.* Washington, DC. http://www.usda.gov/agencies/biotech/ac21/reports/scenarios-4-5-05final.pdf.

U.S. Department of Agriculture. "Agricultural Biotechnology: Frequently Asked Questions. What Is Biotechnology?" http://www.usda.gov/agencies/biotech/faq/html.

U.S. Department of Agriculture. "Glossary of Agricultural Biotechnology Terms." http://www.usda.gov/wps/portal/!ut/p/_s.7_0_A/7_0_1OB?contentidonly=true&navid=AGRICULTURE&contentid=BiotechnologyGlosary.xml. October 2005.

U.S. Department of Agriculture, Animal and Plant Health Inspection Service. "*Changes in Regulation of the Products of Biotechnology by APHIS' Biotechnology Regulatory Services.*" http://www.aphis.usda.gov/brs/eis/index.html. January 2004.

U.S. Department of Agriculture. (2005). Biotechnology, Animal and Plant Health Inspection Service, Regulatory Services. "United States Regulatory Oversight in Biotechnology Responsible Agencies – Overview." http://www.aphis.usda.gov/brs/usregs.html. April 5.

U.S. Environmental Protection Agency. (2000). "*Bacillus thuringiensis* Plant-Pesticides." Biopesticides Registration Action Document: Preliminary Risks and Assessments Sections.

U.S. Environmental Protection Agency. (2003). *A's Regulation of Biotechnology for Use in Pest Management.* http://www.epa.gov/pesticides/biopesticides/reg_of_biotech/eparegof-biotech.htm. June.

U.S. Food and Drug Administration. *Statement of Policy.* (1992). *Foods Derived from New Plant Varieties*, 57 FR 22984. http://vm.cfsan.fda.gov/~lrd/fr92529b.html. May.

U.S. Food and Drug Administration. (2005). Center for Food Safety and Applied Nutrition. *Biotechnology.*

http://www.cfsan.fda.gov/~lrd/biotechm.html. May.

Vencill, W.K. (1996). "Weed Management Systems utilizing Herbicide-resistant Cotton." *Proceedings of the Beltwide Cotton Conferences. Cotton Weed Science Research Conference. v.2.* p. 1532-33.

Verdurme, A. & Viaene, J. (2003). "Consumer Beliefs and Attitude towards Genetically Modified Food: Basis for Segmentation and Implications for Communication." *Agribusiness, 19*(1),*(Winter)*, 91-113.

Virginia Polytechnic Institute & State University. (2005). "Information Systems for Biotechnology, Field Test Releases in the U.S." http://www.isb.vt.edu/cfdocs/fieldtests1.cfm. April 5.

West, G. E., Gendron, C., Larue, B. & Lambert, R. (2002). "Consumers' Valuation of Functional Properties of Foods: Results from a Canada-Wide Survey." *Canadian Journal of Agricultural Economics, 50*(4), *(December)*, 541-58.

Zhong, F., Marchant, M. A., Ding, Y. & Lu, K. (2003). "GM Foods: A Nanjing Case Study of Chinese Consumers' Awareness and Potential Attitudes." *AgBioForum, 5*(4), 136-144.

End Notes

[1] In the United States, under guidelines issued by USDA's Animal and Plant Health Inspection Service (as published in the *Federal Register*, 7CFR340: 340.1), genetic engineering is defined as "the genetic modification of organisms by recombinant DNA techniques" (Fernandez-Cornejo and McBride, 2000). A full biotechnology glossary is in USDA (2005).

[2] Some firms evolved in the 1990s toward developing "life sciences" complexes organized around the development of products such as agricultural chemicals, seeds, foods and food ingredients, and pharmaceuticals based on applications of related research in biotechnology and genetics. However, most of those life sciences companies have since divested their agricultural operations after "failing to realize adequate returns on their investments" (Shoemaker et al., 2003, p.32; Fernandez-Cornejo, 2004, p.42).

[3] Also, there has been an upward trend in the adoption of "stacked gene" varieties (with traits of herbicide tolerance and insect resistance) in the case of cotton and corn.

[4] This yield decrease occurred mostly in early years. HT or Bt genes were introduced into high-yielding cultivars in later years.

[5] Entomologists estimate that the corn rootworm causes up to $1 billion in corn yield losses and insecticide expenditures annually in the U.S. (Comis).

[6] The study used an econometric model that takes into consideration that farmers' adoption of GE crops and pesticide use decisions may be simultaneous and that farmers are not assigned randomly to the two groups (adopters and non- adopters) but that they make the adoption choices themselves. Therefore, adopters and nonadopters may be systematically different. Differences may manifest themselves in farm performance and could be confounded with differences due to adoption. This self- selectivity may bias the results, unless corrected. To account for simultaneity and self-selectivity, the model uses a two-stage econometric model.

[7] In addition, results using an econometric model with the 2001 data showed a small but statistically significant yield increase associated with farmers who adopted Bt corn relative to those using conventional corn varieties. (FernandezCornejo and Li, 2005).

[8] Net returns equal revenues minus variable costs, which include pesticide and seed costs. Seed costs paid by adopters of GE varieties include a technology fee paid by farmers to biotechnology developers and premiums to seed firms.

[9] In addition, using an econometric model with the 2001 data, the ERS study showed a moderate but statistically significant insecticide reduction associated with farmers who adopted Bt corn relative to those using conventional corn varieties (a 4.11-percent decrease in insecticide use was associated with a 10-percent increase in Bt corn adoption).

[10] Conservation tillage includes any tillage and planting system that leaves at least 30 percent of the soil surface covered with crop residue. It includes no-till, ridge-till, and mulch-till (Conservation Technology Information Center, 2004).

[11] The amount that consumers indicate that they are willing to pay for a particular characteristic in a hypothetical situation is sometimes different from the amount that they actually pay when shopping (Lusk, 2003).

[12] In addition, organic foods are available. Use of any GE techniques bars a crop from being certified as organic. Although organic foods still have a small market share (1-2 percent) of total U.S. food sales, their sales have been rising at a rate of 20 percent annually (Dimitri and Greene, 2002).

[13] Consumers may also benefit directly when GE products of the second and third generation are commercialized.

[14] The study estimated the economic gains for various stakeholders associated with adoption by incorporating the potential yield enhancements and savings in pest control costs into models that derive each crop's supply shift resulting from biotechnology. Given domestic and export demands, counterfactual world prices and quantities demanded of the commodities—those that would have prevailed in the market if biotechnology had not been introduced—are determined from market equilibrium conditions. Producer and consumer surpluses in the U.S. and international markets and monopoly profits accruing to the biotech developers and seed firms are then calculated (Price et al., 2003).

[15] Elasticity measures the responsiveness of one economic variable to a change in another (e.g., price and quantity demanded). It is unit free and always expressed in percentage terms.

CHAPTER SOURCES

The following chapters have been previously published:

Chapter 1 – This is an edited, excerpted and augmented edition of a United States Government Accountability Office (GAO), Report to the Committee on Agriculture, Nutrition, and Forestry, U.S. Senate. Publication GAO-09-60, dated November 2008.

Chapter 2 – This is an edited, excerpted and augmented edition of a United States Department of Agriculture, Economic Research Service, Economic Information Bulletin Number 11, dated April 2006.

INDEX

D

E

U

V

W